SpringerBriefs in Food, Health, and Nutrition

Springer Briefs in Food, Health, and Nutrition present concise summaries of cutting edge research and practical applications across a wide range of topics related to the field of food science.

Editor-in-Chief
Richard W. Hartel
University of Wisconsin—Madison, USA

Associate Editors
J. Peter Clark, *Consultant to the Process Industries, USA*
David Rodriguez-Lazaro, *ITACyL, Spain*
David Topping, *CSIRO, Australia*

For further volumes:
http://www.springer.com/series/10203

Avelino Alvarez-Ordóñez • Miguel Prieto

Fourier Transform Infrared Spectroscopy in Food Microbiology

Avelino Alvarez-Ordóñez
Department of Microbiology
University College Cork
Cork, Ireland

Miguel Prieto
Department of Food Hygiene
and Technology
University of León
León, Spain

ISBN 978-1-4614-3812-0   ISBN 978-1-4614-3813-7 (eBook)
DOI 10.1007/978-1-4614-3813-7
Springer New York Heidelberg Dordrecht London

Library of Congress Control Number: 2012937070

Printed on acid-free paper

Springer is part of Springer Science+Business Media (www.springer.com)

Contents

1 Technical and Methodological Aspects of Fourier Transform Infrared Spectroscopy in Food Microbiology Research 1
   1.1 IR Frequency Range and Spectral Windows 1
   1.2 Spectrometer Technology and Spectroscopic Methods 6
   1.3 IR Spectra as a Molecular Fingerprint 9
   1.4 Sample Preparation 12
   1.5 Data Processing and Assessment Approaches 15
   1.6 Databases 17
   1.7 Advantages and Disadvantages of FT-IR Spectroscopy 17

2 Fourier Transform Infrared Spectroscopy to Assist in Taxonomy and Identification of Foodborne Microorganisms 19

3 Fourier Transform Spectroscopy and the Study of the Microbial Response to Stress 23
   3.1 Use of FT-IR Spectroscopy for the Assessment of the Mechanisms of Microbial Inactivation by Food Processing Technologies and Antimicrobial Compounds 24
   3.2 FT-IR Spectroscopy as a Tool for Monitoring the Membrane Properties of Foodborne Microorganisms in Changing Environments 25
   3.3 Detection of Stress-Injured Microorganisms in Food-Related Environments by FT-IR Spectroscopy 27
   3.4 Use of FT-IR Spectroscopy for the Study of Microbial Tolerance Responses 28
   3.5 Applications of FT-IR Spectroscopy for the Study of Spore Composition 29

**4   Fourier Transform Infrared Spectroscopic Methods
    for Microbial Ecology**   31
    4.1   Assessment of Dynamic Changes in Microbial Populations
          by FT-IR Spectroscopy   31
    4.2   Use of FT-IR Spectroscopy to Identify and Quantify
          Microorganisms in Binary Mixed Cultures   32
    4.3   Use of FT-IR Spectroscopy to Detect Food Spoilage
          due to Microbial Activity   33

5   Conclusions and Future Prospects   35

Acknowledgments   45

Abbreviations   47

References   49

Index   55

Chapter 1
Technical and Methodological Aspects of Fourier Transform Infrared Spectroscopy in Food Microbiology Research

Infrared (IR) spectroscopy studies the effect of interaction between matter and radiated energy in the IR range, and this effect is evaluated by measurement of the absorption of various IR frequencies by a sample situated in the path of an IR beam. When a beam of IR radiation is passed through a sample, the radiation can be either absorbed or transmitted, depending on its frequency and the structure of the molecules in the sample. IR radiation excites certain molecular groups, producing vibrations from the excited state at fixed wavelengths. The mechanical behavior of molecules is modified when energy quanta are supplied which change their vibrational and rotational modes. The absorption of energy occurs at frequencies corresponding to the molecular mode of vibration of the corresponding molecule or chemical group. As different functional groups absorb characteristic frequencies of IR radiation, the technique is used in sample identification and investigation of molecular structures.

1.1 IR Frequency Range and Spectral Windows

The electromagnetic spectrum is the range and distribution of radiation according to energy and encompasses a wide range of a seemingly diverse collection of radiant energy, from cosmic rays to x-rays to visible light to IR radiation (Fig. 1.1). Each of them can be considered as a wave or particle traveling at the speed of light and can be defined by the wavelength or wavenumber (Fig. 1.1). The IR region falls between the visible and microwave portions, and the wavenumbers are between 13,000 and 10 cm^{-1} (wavelengths from 0.78 to 1,000 μm), but the most useful vibrational frequencies of most molecules correspond to the mid-IR (MIR) spectrum (between 4,000 and 400 cm^{-1}, or from 2.5 to 25 μm). The IR region is bordered by the visible region at high frequencies and the microwave region at low frequencies. IR absorption measurements are presented as either wavenumbers (v) or wavelengths (λ). Wavenumber is defined by the number of waves per unit length (e.g., number of

A. Alvarez-Ordóñez and M. Prieto, *Fourier Transform Infrared Spectroscopy*
in Food Microbiology, SpringerBriefs in Food, Health, and Nutrition,
DOI 10.1007/978-1-4614-3813-7_1, © Avelino Alvarez-Ordóñez and Miguel Prieto 2012

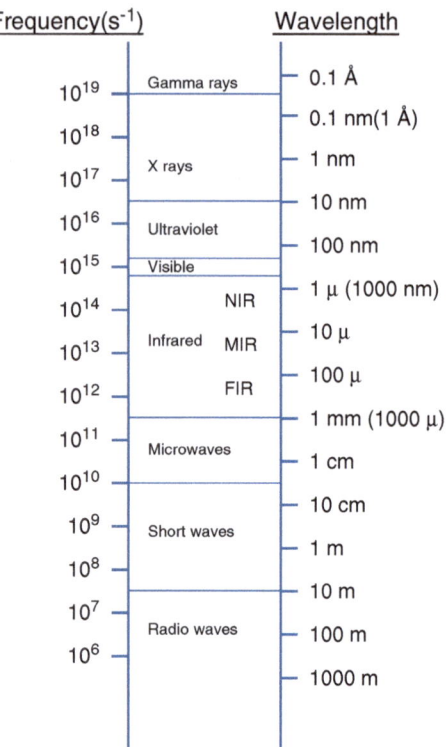

| Frequency(s⁻¹) | | Wavelength |

Fig. 1.1 The electromagnetic spectrum. *IR* infrared, *FIR* far infrared, *MIR* mid infrared, *NIR* near infrared

waves per centimeter, expressed by the unit reciprocal centimeters) and is therefore directly proportional to the frequency, or, understandably, the energy of the IR radiation. Wavelength is defined as the length of one complete wave cycle, or the distance between identical points in the adjacent cycles of a waveform signal, and is usually expressed in the unit micrometer (μm), the recommended unit of wavelength. Use of wavenumber is currently preferred, although IR spectra are sometimes reported in micrometers. Wavenumbers and wavelengths can be interconverted using the following equation:

$$\bar{v}(\text{cm}^{-1}) = \frac{1}{\lambda\,(\mu\text{m})} \times 10^4.$$

The near-IR ($v = 15{,}000$–$4{,}000$ cm⁻¹) and far-IR ($v = 400$–10 cm⁻¹) ranges are not usually employed for purposes such as molecular characterization and identification of bacteria as only overtones (secondary vibrations) and combination vibrations are registered in those regions, and they are therefore difficult to study and interpret from an analytical viewpoint.

Modern IR equipment uses a computer system capable of producing a variety of output spectrum formats, such as transmittance or absorbance versus wavenumber or wavelength. In charts and figures, IR absorption information is normally presented in the structure of a spectrum with wavenumber (or wavelength) as the x-axis and absorption intensity or percent transmittance as the y-axis. Transmittance (T) is the ratio of light transmitted by the sample (I) to the incident light on the sample (I_0) at a specified wavelength. Absorbance (A) is the logarithm to the base 10 of the reciprocal of the transmittance (T):

$$A = \log_{10}(1/T) = -\log_{10} T = -\log_{10} I/I_0.$$

Spectra representing transmittance provide better visual contrast between the intensities of strong and weak bands because transmittance ranges from 0% to 100%, whereas absorbance extends from zero to infinity and is directly proportional to the concentration of a sample at a given wavelength, according to Beer's law.

The atoms in molecules are in continuous motion (at temperatures above absolute zero) with respect to each other. A molecule can have the following types of motions: translation, rotation, and vibration. Translational motion refers to the movement of the molecule through space with a particular velocity. The molecule may rotate around an internal axis, which is a type of oscillation. The molecule may also vibrate since chemically bonded atoms are like masses connected by a bond which can be stretched and compressed, and this particular motion is called stretching. Therefore, a stretching vibration is characterized by a change in the interatomic distance along the axis of the bond between two atoms. The bond angles can vary within certain limits (bending). Thus, bending vibrations are characterized by a change in the angle between two bonds, where the center of mass does not change position and the structure does not rotate.

IR radiation is absorbed by certain molecules and converted into energy of molecular vibration. The chemical bonds that absorb IR radiation vibrate in different ways depending on their own nature. For a molecule to be IR-active, it must undergo a net change in dipole moment as a result of a vibrational or rotational motion. When the frequency of a specific vibration is equal to the frequency of the IR radiation directed at the molecule, the molecule absorbs the radiation. When IR radiation is absorbed by a molecule, the associated energy is converted into various types of vibrational or rotational motions: vibrations can be subdivided into stretching (symmetric and asymmetric) and bending (scissoring, rocking, wagging, and twisting), depending on whether the bond length or angle is changing. Figure 1.2 shows some of the stretching and bending vibrational motions for the H_2O molecule. Each of the vibrational modes has a natural frequency of motion, which is determined by the mass of the atoms connected to each other and the strength of the bond. The larger masses have a lower frequency and the stronger bonds have a higher frequency. The energies associated with rotational transitions are generally much smaller than those from vibrational transitions, and the peaks are found below 300 cm^{-1}. IR spectroscopy deals with transitions between vibrational energy levels in molecules; it is therefore called vibrational spectroscopy. Mostly vibrational modes are detected in IR spectroscopy for biological samples such as solids, liquids, or solutions. Typically, vibrational spectra are measured between 4,000 and 650 cm^{-1} for NaCl optics or between 4,000 and 450 cm^{-1} for KBr optics.

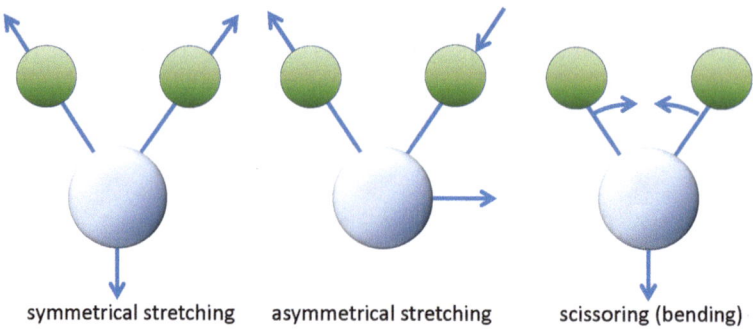

symmetrical stretching   asymmetrical stretching   scissoring (bending)

Fig. 1.2 Stretching and bending vibrational motions for H_2O

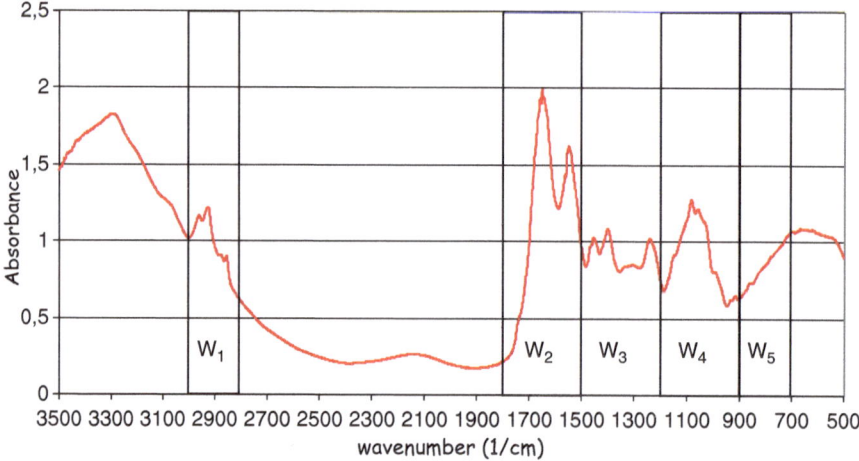

Fig. 1.3 Representative Fourier transform infrared (FT-IR) spectrum (from 3,500 to 500 cm⁻¹) from a bacterial cell sample. The spectrum contains five informative spectral windows: w_1 or fatty acid region (3,000–2,800 cm⁻¹), w_2 or amide region (1,800–1,500 cm⁻¹); w_3 or mixed region (1,500–1,200 cm⁻¹); w_4 or polysaccacharide region (1,200–900 cm⁻¹); and w_5 or fingerprint region (900–700 cm⁻¹)

    As a consequence, the IR spectrum reflects the chemical structure and three-dimensional orientation of the molecules in the sample. IR absorption bands in the MIR spectrum (4,000–400 cm⁻¹), which are mainly due to the vibrational modes, can be tentatively assigned to particular functional groups. Five spectral windows or subranges are commonly depicted within the MIR spectrum that show particular properties (Fig. 1.3):

1. The window between 3,000 and 2,800 cm⁻¹ (w_1), which reflects the dominance of the C–H stretching vibrations of –CH_3 and >CH_2 functional groups, which are abundant in membrane fatty acids, and the side-chain vibrations from some amino acids.

2. The window between 1,800 and 1,500 cm^{-1} (w$_2$), in which the peaks produced by amide I and amide II groups belonging to proteins and peptides are very intense and provide global information on protein structure. The bands near 1,740 cm^{-1} are caused by >C=O stretching vibrations of the ester functional groups in lipids. Also, absorption of nucleic acids occurs in this range owing to >C=O, >C=N, and >C=C< stretching of the DNA or RNA heterocyclic base structures.
3. The window between 1,500 and 1,200 cm^{-1} (w$_3$) is a mixed region influenced by the bending modes of >CH$_2$ and –CH$_3$ groups in proteins, fatty acids, and phosphate-bearing compounds.
4. The window between 1,200 and 900 cm^{-1} (w$_4$), due to the symmetric stretching vibration of PO$_2^-$ groups in nucleic acids and to C–O–C and C–O–P stretching, which reveals the occurrence of carbohydrates and polysaccharides in the cell wall and also the influence of nucleic acids.
5. Finally, the window between 900 and 600 cm^{-1} (w$_5$), which is called the true fingerprint region and contains very specific, weak spectral patterns from ring vibrations of aromatic amino acids (tyrosine, tryptophan, phenylalanine) and nucleotides.

Table 1.1 shows the tentative assignment of some bands frequently found in microbial IR spectra, where frequencies are assigned to functional groups of molecules, allowing the interpretation of IR spectra of particular molecules or structures in biological samples.

Table 1.1 Tentative assignment of some bands frequently found in microbial infrared spectra (peak frequencies have been obtained from the second-derivative spectra) (Adapted from Naumann 2000)

Spectral window	Frequency (cm^{-1})	Assignment
	3,500	O–H stretching of hydroxyl groups
	3,200	N–H stretching (amide A) of proteins
w$_1$	2,955	C–H stretching (asymmetric) of –CH$_3$ in fatty acids
	2,959–2,852	CH, CH$_2$,CH$_3$ in fatty acids
	2,930	C–H stretching (asymmetric) of >CH$_2$
	2,918	C–H stretching (asymmetric) of >CH$_2$ in fatty acids
	2,898	C–H stretching of C–H in methine groups
	2,870	C–H stretching (symmetric) of –CH$_3$
	2,850	C–H stretching (symmetric) of >CH$_2$ in fatty acids
w$_2$	1,740	>C=O stretching of esters
	1,715	>C=O stretching of carbonic acid
	1,680–1,715	>C=O in nucleic acids
	1,695, 1,685, 1,675	Amide I from β-turns of proteins
	1,655–1,637	Amide I bands, of α-helical structures and β-pleated sheet structures
	1,548	Amide II band
	1,550–1,520	Amide II band
	1,515	"Tyrosine" band

(continued)

Table 1.1 (continued)

Spectral window	Frequency (cm^{-1})	Assignment
w$_3$	1,468	C–H deformation of >CH$_2$
	1,400	C=O stretching (symmetric) of COO$^-$
	1,310–1,240	Amide III band components of proteins
w$_4$	1,250–1,220, 1,084–1,088	P=O stretching (asymmetric) of PO$_2$– phosphodiesters
	1,200–900	C–O–C, C–O of ring vibrations of carbohydrates
	1,090–1,085	P=O stretching (symmetric) of >PO$_2$
w$_5$	720	C–H rocking of >CH$_2$
	900–600	"Fingerprint" region

1.2   Spectrometer Technology and Spectroscopic Methods

Two basic types of IR spectrometers are used in analytical techniques for characterization of organic compounds: IR-dispersive spectrometers and Fourier transform IR (FT-IR) spectrometers.

IR radiation (both in dispersive and in FT-IR spectrometers) is obtained from the thermal emission of an appropriate source, the most common radiation source being an inert solid heated electrically in the range from 1,000°C to 1,800°C. There are several types of sources, such as the Nernst glower (constructed of rare-earth oxides in the form of a hollow cylinder), the Globar source (a rod of silicon carbide heated electrically), the carbon dioxide laser, the high-pressure mercury arc, and the Nichrome coil. They all produce continuous radiation, but have different radiation energy profiles and consequently different applications.

Previously used IR-dispersive spectrometers consisted of three basic components: the radiation source, the monochromator, and the detector. This type of IR spectrometer (introduced in the mid-1940s) recorded the amount of energy absorbed by varying the frequency of the IR light using monochromators, and wavenumbers were observed sequentially. The monochromator was used to limit and disperse a broad range of IR radiation and provide a calibrated series of electromagnetic energy bands of determinable wavelength range by using prisms or gratings as the dispersive components together with input and output slits, opening mechanisms, mirrors, and filters. Because of that, they are named dispersive spectrometers.

FT-IR spectrometers were developed for commercial use in the 1960s, but they did not find widespread use in laboratories until recently because of initial disadvantages such as high component costs and lack of versatility. Several circumstances have contributed to the successful development of FT-IR spectroscopy. FT-IR spectroscopes currently in use include a Michelson interferometer, which is technically superior to conventional monochromators used in dispersive spectroscopy. Besides, the fast Fourier transform algorithm is able to compute the discrete Fourier transform and its inverse, so the raw signal is quickly converted into a recognizable

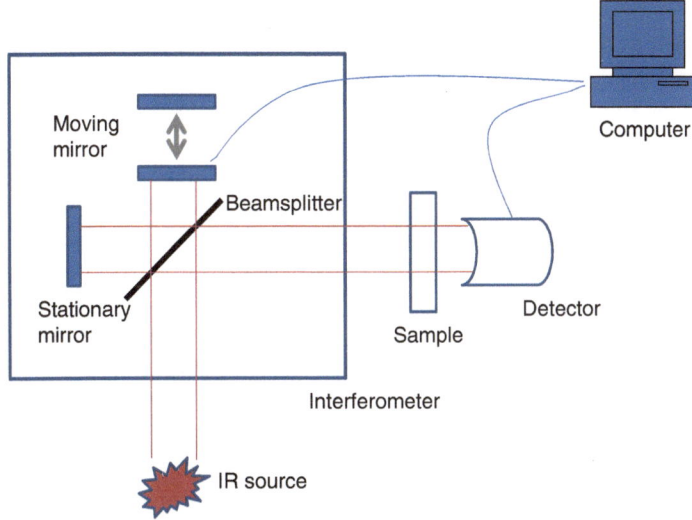

Fig. 1.4 Basic components of an FT-IR spectrometer, including the light source, the detector and a Michelson interferometer consisting of a beamsplitter, a fixed mirror, and a moving mirror

absorbance spectrum. Other additional advantages such as superior speed and sensitivity shown by FT-IR spectroscopes have resulted in the gradual replacement of the dispersive instruments for most applications. The use of chemometric tools, which allow the extraction of qualitative and quantitative information from the spectra, and the versatility of the spectroscopic technique in new applications, which permits the analysis of samples in different conditions (liquid, suspended, powered, or dehydrated), should also be mentioned.

Modern FT-IR systems are made up of three components: the radiation source, the interferometer (instead of a monochromator), and the detector. Figure 1.4 shows the basic components of an FT-IR spectrometer, including the light source, the detector and a Michelson interferometer (which consists of a moving mirror, a fixed mirror, and a beamsplitter). The interferometer interferometrically modulates radiation by splitting a beam of light into two paths so that one beam strikes a fixed mirror and the other a movable mirror. When the reflected beams bounce back and are recombined, an interference pattern results. The interferometer produces interference signals, which contain IR spectral information generated after a beam has passed through a sample. Repetitive interference signals are produced and measured as a function of the optical path difference by a detector.

In modern FT-IR spectrometers, the IR beam is guided through an interferometer where spectral encoding takes place. The interferometer performs signal encoding and converts the initial frequencies into a special form decipherable by the detector. Instead of examining each wavenumber in succession, as in dispersive IR spectroscopy, all frequencies are examined simultaneously. After the beam has passed through the sample or has been reflected, the measured signal is the resulting interferogram, which contains information from the entire IR region. Finally, the beam passes to the detector and the measured signal is digitalized and sent to the computer, where Fourier transformation of the signal occurs. This Fourier transformation of the signal can be viewed as a mathematical method of converting the individual frequencies from the interferogram, by using a mathematical procedure able to transform a function from the time domain to the frequency domain. The final IR spectrum is then depicted on the screen and eventually saved to the disk, and is identical to that obtained from conventional (dispersive) IR spectroscopy.

The response times of many detectors used in dispersive IR spectrometers are too slow and cannot adjust to the rapid scanning times needed for FT-IR interferometers. There are two popular detectors which are included in most FT-IR spectrometers: deuterated triglycine sulfate (DTGS) and mercury cadmium telluride (MCT). For rapid-scanning interferometers, MCT detectors are used since they exhibit very fast responses, whereas for slower scanning types, DTGS detectors can be used. The DTGS detector is a pyroelectric detector that operates by sensing the changes in temperature of an absorbing material. The MCT detector is a photon (or quantum) detector that depends on the quantum nature of radiation, or the interaction between radiation and the electrons in a solid. The MCT detector is more sensitive than the DTGS detector, but usually it has to be cooled (77 K) using liquid nitrogen to cut out thermal noise, whereas DTGS detectors operate at room temperature.

Several spectroscopic techniques have been used to obtain IR spectra for characterization of the molecular composition of foodborne bacteria (Fig. 1.5). In FT-IR transmittance techniques, the sample is placed in the path of the IR beam and scanned (Fig. 1.5a). Solid samples can be finely ground with potassium bromide (a matrix transparent in the MIR region) and mechanically pressed to form a hard, translucent pellet, with a sample concentration in KBr in the range from 0.2% to 1%. Liquid samples with a cell suspension can be deposited on an appropriate window and stove-dried before being measured. For wet samples (most biological samples), water-insoluble, IR-transparent optical windows are used (CaF_2, BaF_2, ZnSe, ZnS, or germanium). In diffuse reflectance FT-IR (DRIFT) spectroscopy, the IR beam is projected into the sample, where it is reflected by, scattered by, and transmitted through the sample. The part of the IR light that is diffusely scattered within a sample and returned to the detector optics is considered to be diffuse reflection (Fig 1.5b). High-scattering samples such as freeze-dried biological samples can be analyzed. In attenuated total reflectance (ATR) spectroscopy, the sample is placed onto an optically dense crystal of relatively high-refractive index, needing little or no sample preparation (Fig. 1.5c). The IR beam is reflected from the internal surface of the crystal and creates an evanescent wave, which extends beyond the surface of the crystal and projects into the sample in close contact with the ATR crystal. Some of the energy of the evanescent wave is absorbed by the sample, and the reflected radiation is passed to the detector in the spectrometer as it exits the crystal. For the DRIFT and ATR FT-IR methods, some studies (Winder and Goodacre 2004) have focused on the differences in the discrimination capacity and the information obtained when analyzing bacteria, and apparently the ATR FT-IR method (which yields a biochemical profile of the surface chemistry of cells) achieved a better subspecies differentiation than the DRIFT method. Besides the limited sample

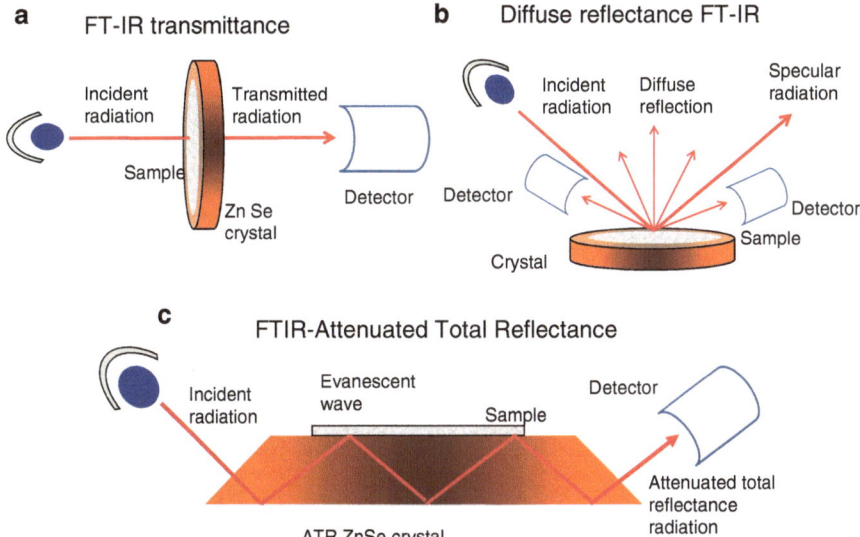

Fig. 1.5 The main spectroscopic methods used to characterize molecular composition and stress response in foodborne pathogenic bacteria by FT-IR spectroscopy: (**a**) transmittance FT-IR; (**b**) diffuse reflectance FT-IR; (**c**) attenuated total reflectance (*ATR*) FT-IR. (Reprinted from Journal of Microbiological Methods, 84, 369–378, Álvarez-Ordóñez, A.; Mouwen, D.J.M.; Lopez, M.; Prieto, M. Use of Fourier transform infrared spectroscopy as a tool to characterize molecular composition and stress response in foodborne pathogenic bacteria, Copyright (2011), with permission from Elsevier)

preparation (liquid samples can be processed), another advantage is the availability of specially designed ATR cuvettes suitable for the measurement of multiple samples. Finally, FT-IR microspectroscopy is a new technique which, coupling an FT-IR spectrometer and a microscope, is able to inspect particular areas on a surface such as an agar plate, and to obtain reflectance or transmittance spectra from samples comprising a few hundred cells, for example, microcolonies grown after 6–10-h. For this purpose, isolation and purification of the microorganisms to be measured are not necessary, provided that the microcolonies are well separated on the plate.

1.3   IR Spectra as a Molecular Fingerprint

A complex, fingerprint-like resonance absorption band is obtained when a substance is radiated with a continuous spectrum of IR light, and the intensity of the absorption bands stems from scanning before and after passing the IR beam through the substance. FT-IR spectra obtained from pure compounds are very characteristic and they can be considered "fingerprints" or molecular patterns distinctive of a particular bacterial strain. The quantity and quality of information obtained from organic compounds is very high; inorganic compounds are usually much simpler than organic compounds and thus the quantity of information obtained from them is usually much less. In complex samples such as bacterial biomass, the IR spectrum is composed of unique broad and complex contours instead of isolated peaks,

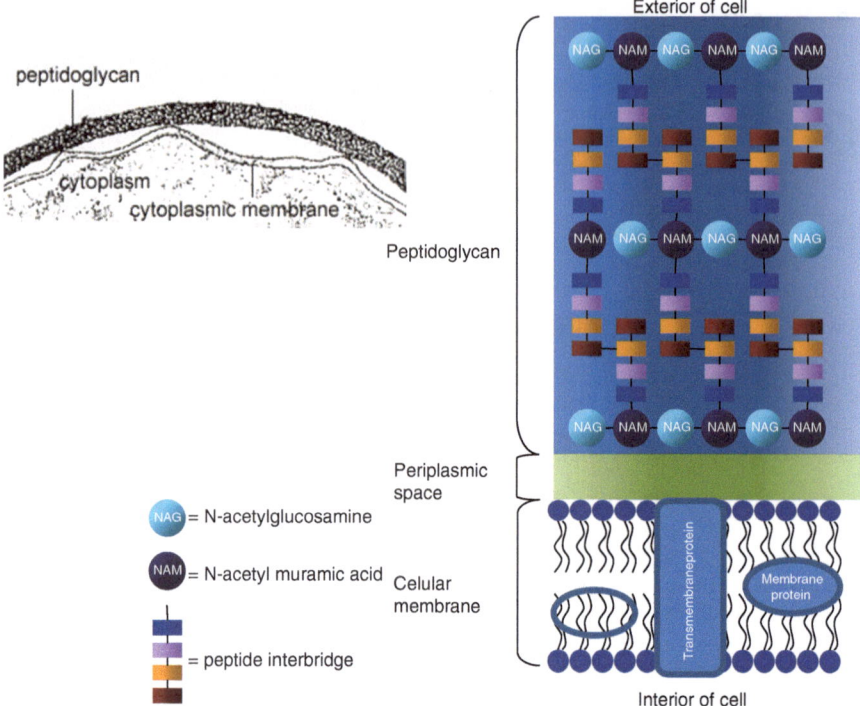

Fig. 1.6 A typical Gram-positive cell, showing the arrangement of the peptidoglycan (with blocks of *N*-acetylglucosamine and *N*-acetylmuramic acid), the periplasmic space, and the cell membrane

representing the superposition of all IR vibrational modes of the molecules in the sample. As the IR spectrum reflects the global chemical composition of the sample, it can afford information on existing taxonomic differences, or on chemical changes undergone owing to stressful environments, and the IR spectrum can be used in the identification, characterization, and quantification of the sample.

Several researchers have conducted experiments to assign the main cell structures or biological blocks—such as the cell wall or molecules such as lipopolysaccharides (LPS) and outer-membrane proteins—to the IR bands and consequently to the clustering observed (Kim et al. 2005, 2006, 2011). Other authors have produced classification schemes for several genera of lactic acid bacteria that agree with a taxonomic classification based on the composition of the cell wall (Mouwen et al. 2011). There is significant evidence that the cell surface or its components are major elements with a strong effect on the IR spectrum.

The cell wall structure has physiological and mechanical functions and prevents bacteria from bursting from high internal osmotic pressure. The differences in the cell envelop between Gram-positive and Gram-negative bacteria are clearly outlined (Figs. 1.6 and 1.7). Gram-positive cell walls contain a thick and rigid layer of

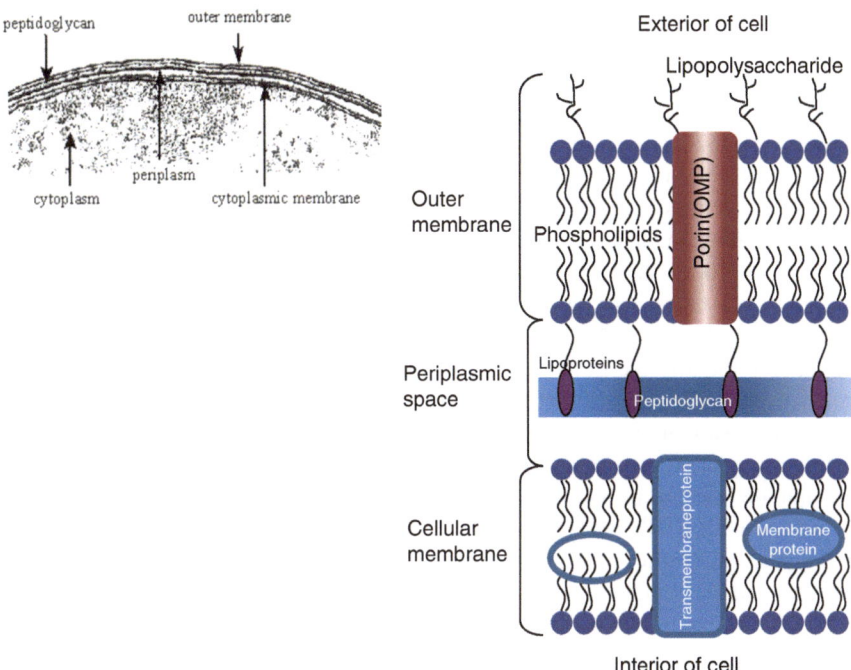

Fig. 1.7 A typical Gram-negative bacterium, with the peptidoglycan sandwiched between the outer membrane and the cell membrane. *OMP* outer-membrane protein

peptidoglycan covering the cytoplasmic membrane, and may also have an S-layer and a slime capsule on top of the peptidoglycan. Gram-negative cell walls contain a thin layer of peptidoglycan between the cytoplasmic membrane and the outer membrane. The peptidoglycan consists of linear polysaccharide chains of alternating *N*-acetylglucosamine and *N*-acetylmuramic acid residues linked by $\beta(1 \rightarrow 4)$ glycosidic bonds, cross-linked by short peptides (pentapeptides or tetrapeptides) whose assembly differs between different species of bacteria. The most common cross-bridge in Gram-negative bacteria is composed of D-alanine, L-alanine, D-glutamic acid, and diaminopimelic acid. In Gram-positive bacteria (*Staphylococcus*), it consists of L-alanine, L-lysine, D-glutamine, and D-alanine. The Gram-positive cell walls also contain teichoic acids that are covalently bound to the peptidoglycan and serve as a link tying the different layers together. The peptidoglycan is also present in Gram-negative bacteria, but is much thinner (representing 15–20% of the cell wall), does not contain teichoic acids, and is only intermittently cross-linked by means of lipoproteins covalently bound to the peptidoglycan in the cell walls. In Gram-negative bacteria, there is an outer membrane outside the peptidoglycan which contains phospholipids in the inner layer and LPS in the outer layer. LPS is composed of three regions: a lipid region called lipid A (which anchors the LPS to the membrane), the inner and outer core oligosaccharides, which contain ketodeoxyoctonate, heptose, glucose, and glucosamine sugars, and a highly variable O-specific side chain (a polysaccharide with antigenic capacity which is used in bacteria typing). Other structures that may be present in the bacterial envelop are the capsule and the flagella. There are significant differences in the type of sugars and the organization of LPS and other antigenic structures between genera and species of bacteria.

The differences in the type of sugars and the organization of LPS and other anti-genic structures in bacteria can be detected using FT-IR and chemometric analysis. Kim et al. (2005) showed that the FT-IR technique is very useful for taxonomic investigations because of its discriminatory capacity as compared with deoxycholic acid polyacrylamide gel electrophoresis in LPS analysis. Nonetheless, they were not able to identify the LPS structure of the *Salmonella* serotypes analyzed which could explain the spectral differences observed between the different *Salmonella* serotypes. Similar results were obtained when analyzing *Salmonella* outer-membrane protein profiles from several serotypes by sodium dodecyl sulfate poly-acrylamide gel electrophoresis, which was inferior in discriminating power as compared with FT-IR analysis, which achieved 100% correct classification of the serotypes. For Gram-positive bacteria, a close correlation between FT-IR spectral features and the composition of the cell wall has been found. According to Mouwen et al. (2011), FT-IR grouping of lactic acid strains revealed the composition of the cell wall (Fig. 1.8a). Clusters were defined on the basis of the absence or presence of teichoic acid in the cell wall and its carbohydrate (glycerol or ribitol). Another feature revealed was that the classification was also in accordance with the type of amino acids present in the peptide bridge. Other authors have reported that the occurrence of certain molecules in the cell wall of lactic acid bacteria can be revealed in the FT-IR spectrum (Curk et al. 1994). In conclusion, FT-IR spectroscopy can afford information additional to phenotypic and genotypic data which may help to establish a more robust taxonomic classification. The IR spectrum constitutes an image of the overall chemical composition, although compounds and structures, such as the cell wall, have a major influence on the spectrum.

1.4 Sample Preparation

A basic stage in FT-IR analysis of microbial biomass is sample preparation. The IR spectrum is an expression of the composition of the sample (cell biomass), and anything that can modify the composition of the cell structures will be reflected in the final IR spectrum. There are many sources of variability during sample

---

Fig. 1.8 (continued) (**b**) Differentiation of 16 lactic acid bacteria belonging to the genera *Lactobacillus, Carnobacterium*, and *Weisella* (3 replicates from 16 reference strains) using the scores of the first and second canonical dimensions from the canonical discriminate analysis of their first-derivative IR spectra in the 1,200–900-cm^{-1} range. (Reprinted from Vibrational Spectroscopy, 56, 193–201, Mouwen, D.J.M.; Hörman, A.; Korkeala, H.; Alvarez-Ordoñez, A.; Prieto, M. Applying Fourier transform infrared (FTIR) spectroscopy and chemometrics to the characterization and identification of lactic acid bacteria, Copyright (2011), with permission from Elsevier)

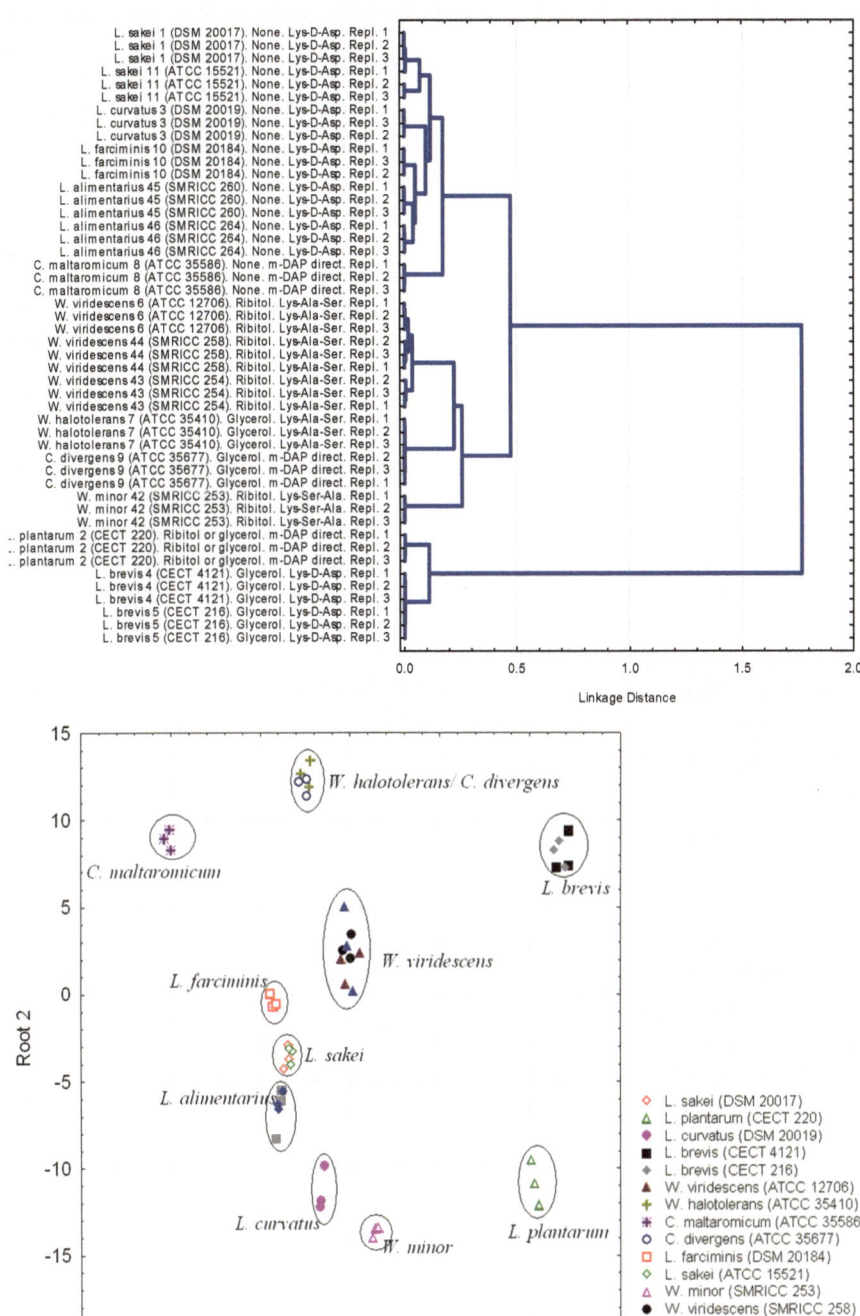

Fig. 1.8 (a) Dendrogram obtained from first-derivative IR spectra in the 1,200–900-cm⁻¹ range of lactic acid bacteria belonging to the genera *Lactobacillus*, *Carnobacterium*, and *Weisella* (three replicates from 16 reference strains), with cluster analysis having been performed with Pearson product moment correlation coefficient and the Ward algorithm clustering method.

preparation which should be minimized. Factors such as the formulation of the growth culture medium and the growth phase (lag, logarithmic, stationary phase) of the microbial population have been shown to alter the final composition of the cells and consequently the IR spectrum. The selection and preparation of the sample is also essential for those techniques which involve a grinding and mixing step. Therefore, to obtain reproducible spectral data, a standardized experimental protocol in relation to medium preparation, incubation time and temperature, cell harvesting conditions, sample preparation, and FT-IR measurement should be carefully followed. Other important reasons to achieve a high degree of standardization arise from the need to exchange spectral data between laboratories and to generate validated reference libraries including FT-IR spectra of bacteria isolated and cultivated in different laboratories.

Variability between IR spectra even when several samples obtained from a single strain are repetitively measured occurs, and it is accordingly pertinent to determine the reproducibility of IR spectra for a given strain under the same or different preparation conditions. Standardized experimental protocols including data acquisition and evaluation procedures have been published, and the results show that reproducible results can be obtained provided careful attention is paid to protocols (Helm et al. 1991a, b; Naumann 2000).

For transmittance FT-IR spectroscopy, samples are handled in two ways. In the first procedure, strains are cultured either in broth or in solid medium in standardized cultivation conditions, paying special attention to the growth medium (batch), temperature, and incubation time. A sufficient amount of biomass (about 10–60 μg dry weight) is carefully harvested from the agar plates and washed several times to eliminate possible impurities. Finally, it is suspended in 100 μl distilled sterile water, placed (20 μl) in a ZnSe (zinc selenide) window, and stove-dried (5 min, 60°C). Alternatively, the strains can be incubated in culture broth and harvested by centrifugation (8,000g for 5 min at 4°C) and suspended in 100 μl phosphate-buffered saline. After purification using phosphate-buffered saline, the wet biomass is placed (10 μl) in a dense, high-refractive-index crystal such as zinc selenide, thallium bromide–thallium iodide (KRS-5), or germanium, and dried in different ways (e.g., in a stove for about 15 min at 50°C). For some samples in solution, a concentrated solution of the sample in a suitable solvent (e.g., CH_2Cl_2) can be prepared. The solution is transferred with a pipet onto the IR plates.

Another procedure, which is more time-consuming, is the preparation of pellets or disks by mixing the sample and an optically active substance, such as KBr. Solid samples that are difficult to dissolve in any suitable IR-transmitting solvents are pelletized. The sample (0.5–1.0 mg) is finely ground and intimately mixed with approximately 100–200 mg of dry KBr powder using an agate stone mortar and pestle, or a vibrating ball mill. Samples may be lyophilized before being processed. The mixture is then pressed into a transparent disk using a device such as a hydraulic pellet press in an evacuable die at sufficiently high pressure (10–20 t). To minimize band distortion due to scattering of radiation, the sample should be ground to particles of 2 μm (the low end of the radiation wavelength) or less in size. Pellets should be dehydrated and kept dry (e.g., in the presence of silica gel), otherwise the IR spectra produced by the pellet technique may exhibit bands at 3,450 and 1,640 cm^{-1} due to absorbed moisture.

Whereas measurement by FT-IR macrospectroscopy takes 2 days, requiring an additional purification step and the choice of a single colony to perform the analysis, analysis by FT-IR microspectroscopy uses a specific step: organisms are grown to microcolonies and directly transferred from the agar plate to the IR-transparent ZnSe carrier. This can be done using a stamping device, which results in spatially accurate replicas. The samples have to be dried as well.

1.5   Data Processing and Assessment Approaches

For most common single compounds, the IR spectrum can be identified by comparison with a library of known compounds. For biological samples such as microbial biomass and because of the complexity of the IR spectrum, an appropriate mathematical transformation should be done first. Before spectra can be studied, they should be transformed to minimize variability due to methodological conditions and at the same time amplify the chemically based spectral differences. A series of preliminary transformations are usually performed, such as baseline correction, normalization, smoothing, and derivatization (first or second derivative commonly using the Savitzky–Golay algorithm).

Baseline correction achieves flatter baselines and averages the baseline to zero, which eliminates the dissimilarities between spectra due to shifts in the baseline. Smoothing reduces noise in the signal, and increases the information content of the IR spectrum. Derivatization is used to improve resolution and correct baseline variability. Other applications are meant to decrease replicate variability, resolve overlapping peaks, and amplify spectral variations. Both first- and second-derivative transformations are commonly employed using the Savitzky–Golay algorithm. First-derivative transformation is generally used for the taxonomic classification of bacteria, and it is sufficient when the differences in the IR spectra are significant enough and visible to the naked eye. Second-derivative spectra increase the number of discriminative features associated with bacterial spectra and improve the clarity, with the net effect of an increase in spectral resolution. Second-derivative spectra are commonly used for bacterial classification, studying the growth phase of bacteria, and detecting sublethal injury of bacteria due to heat, chlorine, radicals, and sonication. Normalization of spectra is also a prerequisite for advanced statistical analysis of bacterial spectra. This transformation tries to minimize differences due to sample amount, eliminates the path length variation, and also reduces the differences between each single measurement of the same sample, which is usually applied to study the spectra of bacteria at different growth phases. The spectra are normalized to the most intense band or to the same integrated intensity in a given spectral region. Usually amide I band is used as an internal standard for normalization.

It is usually achieved by setting the absorption at $1,800$ cm^{-1} to 0 and the maximal absorption, located around $1,650$ cm^{-1}, to 1.

Once the spectra have been transformed, they can be mathematically analyzed and compared using data reduction tools, regression techniques, and classification methods. To obtain a relevant picture of the relationships between spectra and assist in the classification process, several chemometric procedures can be used. Two general approaches, i.e., supervised and unsupervised techniques, can be followed depending on the previous knowledge of elements (IR spectra) and relationships. Supervised techniques make use of a priori knowledge of classes to guide the characterization or classification process. Unsupervised methods try to disclose naturally occurring groups and structures within the data set without previous knowledge of class assignment. Only a brief description of the most widely used techniques will be presented as an exhaustive and detailed description beyond the scope of this review.

There are many unsupervised techniques, such as cluster analysis, factor analysis, and principal component analysis. Principal component analysis is used as a data reduction tool which generates a new set of noncorrelated variables for each of which a score (or value) for each sample is calculated. It is a useful tool in classification as the graphical display of scores may disclose clustering patterns. Cluster analysis arranges unknown elements into groups (so-called clusters) on the basis of their resemblance, expressed by means of different mathematical distance or similarity coefficients. Factor analysis is a multivariate procedure whose main objective is to reduce the dimensionality of a data set and to detect hidden relationships between variables. A set of correlated variables is thus transformed to a set of uncorrelated, hidden variables (factors) ranked by variability in descending order.

Techniques such as soft independent modeling of class analogy, linear discriminant analysis, canonical variate analysis, and artificial neural networks are examples of supervised methods, where an ample collection of well-defined spectra is used to develop a function or train a model which will be used later for classification and identification. Soft independent modeling of class analogy is a statistical method for supervised classification of data which builds a distinct confidence region around each class after applying principal component analysis. New measurements are projected in each principal component space that describes a certain class to evaluate whether they belong to it or not. Linear discriminant analysis is a supervised learning technique for classification. The aim is to find a linear combination of attributes (function) which characterize or separate two or more classes. Canonical variate analysis separates objects (samples) into classes by minimizing the within-class variance and maximizing the between-class variance. Artificial neural networks are mathematical models working in a supervised manner, that can be used in identification owing to their capacity to generalize, find patterns in data, and model complex relationships between inputs and outputs. Regression techniques such as principal component regression and partial least squares regression are used for prediction and quantification purposes. Both are linear regression methods that create components or factors as new independent explanatory variables in a regression model.

1.6 Databases

Construction of FT-IR databases has been a long-term objective as routine identification of foodborne bacteria can only be performed if an extensive spectral database for most pathogens is established. This objective has been pursued by many research groups but only few of them have been able to develop broad libraries with a commercial purpose. FT-IR systems for detection, quantification, and differentiation of microorganisms with a high degree of automation are now commercially available. Among the laboratories that have developed and validated databases for microbial identification, the Institute for Microbiology of the Zentralinstitut für Ernährungs- und Lebensmittelforschung should be mentioned, with libraries for several microbial groups (yeasts, coryneform bacteria, bacilli, micrococci, staphylococci, bifidobacteria, clostridia, pseudomonads, lactobacilli, lactococci, *Acetobacteraceae*).

1.7 Advantages and Disadvantages of FT-IR Spectroscopy

A comprehensive list of the advantages and disadvantages of FT-IR spectroscopy has been given by Naumann (2000) and Davis and Mauer (2010).

It is possible to obtain quality data quickly and, in principle, with little expertise. Within a few minutes of removing single colonies from a microbial cell plate, results are obtained. The routine operation is inexpensive compared with genotaxonomic systems for bacterial identification, although the equipment is costly. The sample is not destroyed (in most of the techniques) as the bacterial biomass stays intact and can be analyzed further, allowing typing or DNA extraction. The preparation of the sample is minimal and very little sample is needed (less than a few milligrams). Although many commercial spectrometer models are available, they show a certain degree of standardization in regard to data acquisition and data reproducibility. Proprietary software tools are also readily available and can be easily implemented in routine analysis. There is also software for spectral mathematical transformation and analysis.

Nowadays, there are several spectroscopic methods (transmittance, reflectance) which are able to analyze multiple preparations as samples in the form of liquid, gas, powder, solid, or film. Experience shows that the technique has many different applications in food microbiology, such as identification and discrimination of bacteria, stress response, and the like. Detection of specific cell components such as storage materials, spores, bacterial capsules, and properties such as drug resistance and cell–drug interaction (including its monitoring and characterization) is also possible. Bacteria in several physiological forms (biofilms, planktonic cells) can be analyzed. IR spectroscopy has also shown good complementation and reasonable agreement with other taxonomic tools, such as several types of genotyping schemes developed for various microbial groups. Classification can be performed at very different taxonomic levels, which makes IR spectroscopy useful at the genus, species, and even strain level, although sometimes classifications obtained at the genus level are not taxonomically relevant in all cases, showing discrepancies with other standard methods based on genotypic data. A developing field is the application of FT-IR spectroscopy in quantitative analysis: IR spectra not only provide information on the bacterial cell composition but are also able to quantify the number of bacteria or number of functional groups present in a sample.

The greatest disadvantages are related to the necessity of obtaining sufficient biomass. It is implicit that the method can be only applied to microorganisms that can be cultivated where sufficient amount of biomass can be collected. This fact implies that fastidious microorganisms, noncultivable species, or those with particular growth requirements may present problems for retrieval of the IR spectrum. There is a significant influence of environmental conditions which can alter the IR spectra, and this has long been recognized. As mentioned earlier, special care should be taken with regard to the sample preparation and especially in the preliminary steps, such as culture medium preparation, incubation temperature and time, and cell harvesting conditions. Only pure cultures can be analyzed since mixed biomass would yield overlapping spectra, and erroneous results in identification or characterization. Nonetheless, researchers are obtaining good results in the analysis of mixed cultures, and it is also possible to investigate mixed cultures using microspectroscopy provided single colony growth is achieved. Similarly to other identification systems, extensive and high-quality databases or libraries encompassing large taxonomic entities should be developed first since they are needed for reliable identifications. Only with appropriate mathematical protocols (e.g., artificial neural networks) is it possible to detect outliers and avoid errors in identification. Although the acquisition of IR spectra seems to be an easy task, selection of the spectrum and analysis should be rigorous and needs chemometric expertise.

Chapter 2
Fourier Transform Infrared Spectroscopy to Assist in Taxonomy and Identification of Foodborne Microorganisms

The most extensive application of Fourier transform infrared (FT-IR) spectroscopy in food microbiology has been the identification and characterization of bacteria, yeast, fungi, and algae. For identification, the infrared (IR) spectrum of an unknown species is compared with all spectra present in a previously constructed database and is matched to the reference strain whose spectrum is most similar. The IR spectrum of bacterial biomass was recorded and studied as early as the 1950s (Stevenson and Bolduan 1952; Thomas and Greenstreet 1954; Riddle et al. 1956; Goulden and Sharpe 1958). The IR spectra of several lactic acid bacteria were reported to show visually appreciable differences (Goulden and Sharpe 1958). Nonetheless, the development of the technique did not occur until recently and a review on this subject at that time (Norris 1959) summarized that "although bacteria definitely exhibit IR spectra that are unique for individual strains, the identification of bacteria via IR techniques is impractical due to inherent disadvantages."

Identification and classification of bacteria using FT-IR spectroscopy is based on the fact that the IR spectra are a reflection of the overall molecular composition. Strains that differ in their molecular makeup will therefore show distinct vibrational spectra. The capacity of FT-IR spectroscopy to identify and characterize unknowns is reflected by studies comparing the spectroscopic method with phenotypic and genotypic identification methods. Several studies have highlighted the capacity to differentiate strains using immunologic or genetic techniques. FT-IR spectroscopy was able to differentiate strains belonging to *Serratia marcenscens*, and this differentiation was consistent with the macrorestriction results using multilocus enzyme electrophoresis (Irmscher et al. 1999). Strains belonging to *Campylobacter jejuni* and *C. coli* could be separated according to types obtained by means of a genotyping method such as the enterobacterial repetitive intergenic consensus polymerase chain reaction (Fig. 2.1) (Mouwen et al. 2005, 2006). Differentiation of *Yersinia enterocolitica* isolates at the species and subspecies levels was successfully achieved, and they were separated according to their biotype/serotype into the main biotypes (biotypes 1A, 2, and 4) and serotypes (serotypes O:3, O:5, O:9, and "non-O:3, non-O:5, and non-O:9") (Kuhm et al. 2009). Several articles have shown the capacity of

A. Alvarez-Ordóñez and M. Prieto, *Fourier Transform Infrared Spectroscopy in Food Microbiology*, SpringerBriefs in Food, Health, and Nutrition, DOI 10.1007/978-1-4614-3813-7_2, © Avelino Alvarez-Ordóñez and Miguel Prieto 2012

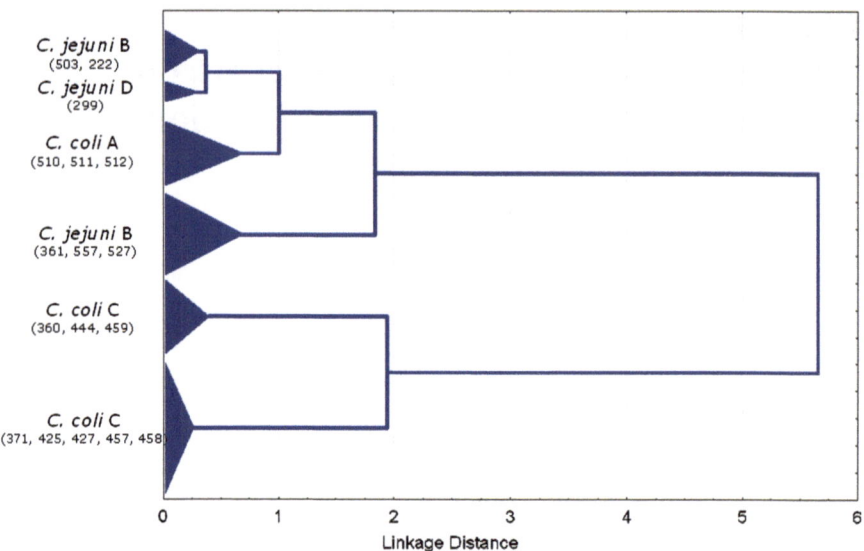

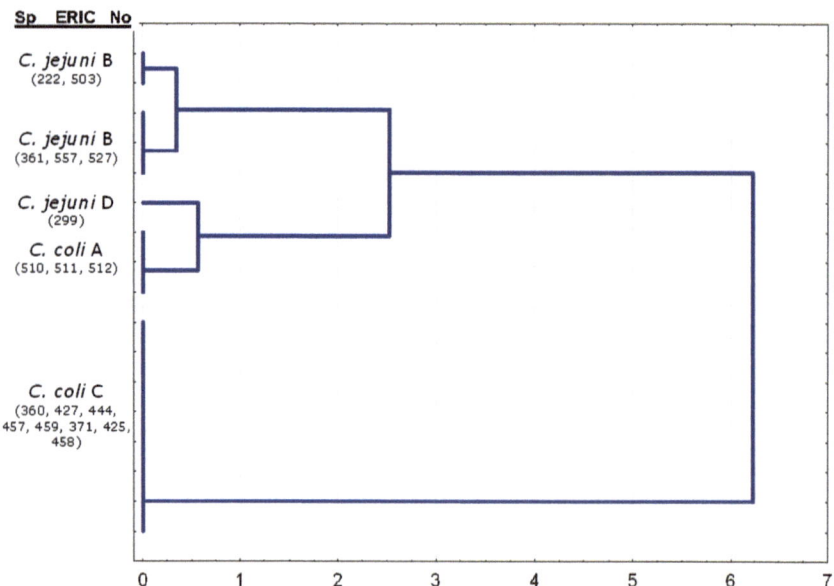

Fig. 2.1 (**a**) Dendrogram obtained from FT-IR spectral data of thermophilic *Campylobacter* strains, with cluster analysis having been performed with the Pearson product moment correlation coefficient and the Ward algorithm method. Three replicates (48-h-old culture) measured independently belonging to 17 strains were included in the analysis (30 wavelengths from the w_4 spectral window). (**b**) Dendrogram obtained from enterobacterial repetitive intergenic consensus (*ERIC*) polymerase chain reaction profiles of 17 *Campylobacter* strains, with cluster analysis having been performed with the Sørensen–Dice coefficient and with the Ward algorithm method. (Copyright © American Society for Microbiology, Applied and Environmental Microbiology, 71, 4318–4324)

FT-IR spectroscopy to differentiate between *Salmonella* phagotypes or serovars (Seltmann et al. 1994; Kim et al. 2005, 2006; Baldauf et al. 2006; Preisner et al. 2010). Other pathogens that have been studied and successfully discriminated at subspecies level (intraspecific diversity) are enteropathogenic *Escherichia coli* (Gilbert et al. 2009), *Bacillus cereus* (Lin et al. 1998; Al-Holy et al. 2006; Mietke et al. 2010; Kim et al. 2011), *Brucella* (Miguel Gomez et al. 2003), *Cronobacter sakazakii* (Lin et al. 2010), *Listeria monocytogenes* (Holt et al. 1995; Lefier et al. 1997; Rebuffo-Scheer et al. 2007; Janbu et al. 2008; Davis and Mauer 2011), and *Mycobacterium bovis* (Winder et al. 2006).

Particular applications with the aim of distinguishing specific groups within a genus have also been described. A complete discrimination of group I and group II *Clostridium botulinum* strains was achieved (Kirkwood et al. 2006). *Staphylococcus aureus* strains isolated from raw milk and raw milk cheeses were identified and separated from other species of *Staphylococcus* (Lamprell et al. 2006). Coagulase-negative staphylococci were distinguished from *S. aureus*, and glycopeptide-intermediate *S. aureus* were distinguished from glycopeptide-susceptible methicillin-resistant *S. aureus* (Amiali et al. 2007, 2008).

Most of the work has been devoted to discriminating foodborne pathogens, although a significant number of articles have dealt with spoilage or beneficial bacteria. Several authors have successfully characterized isolates of lactic acid bacteria obtained from food products, such as milk (Weinrichter et al. 2001), cheese (Amiel et al. 2000; Oust et al. 2004; Savic et al. 2008), kefir (Bosch et al. 2006), and meat products (Oust et al. 2004; Mouwen et al. 2011). Spores belonging to different species of *Bacillus* (*B. atrophaeus*, *B. brevis*, *B. mycoides*, *B. circulans*, *B. lentus*, *B. megaterium*, *B. subtilis*, *B. thuringiensis*) present in several environments including food can be easily differentiated (Beattie et al. 1998; Filip et al. 2004; Brandes-Ammann and Brandl 2011). Finally, besides bacteria, foodborne yeasts (Kummerle et al. 1998; Paramithiotis et al. 2000; Wenning et al. 2002; Nieuwoudt et al. 2006) and filamentous fungi (Fischer et al. 2006) have also been studied.

Chapter 3
Fourier Transform Spectroscopy and the Study of the Microbial Response to Stress

A number of research articles have assessed the structural modifications occurring in food-associated microorganisms in response to environmental stress conditions. Some of these articles have studied in depth the mechanisms of death induction resulting from vegetative cell exposure to different food processing technologies, antimicrobial compounds, and adverse environmental conditions, with most of the studies focused on their effects on the cytoplasmic membrane composition and structure. The successful application of Fourier transform (FT-IR) spectroscopy in this field has led to more ambitious studies demonstrating the capacity to detect and quantify injured vegetative cells in food products and to check the efficacy of food processing treatments. Furthermore, several authors have recently shown FT-IR spectroscopy is also a suitable method to evaluate stress-induced changes in spore components, suggesting the use of this technology to monitor the efficacy of sterilization techniques in inactivation of spore-forming microorganisms. Other research articles have emphasized the ability of FT-IR spectroscopy to study dynamic changes in bacterial populations and to discriminate between different phenotypes of a given bacterial strain. This offers the possibility of identifying phenotypes relevant for food safety, i.e., those showing an extremely high resistance to food processing systems and harsh environments, such as the phenotypes resulting from bacterial adaptive tolerance responses.

The following sections provide an overview of the novel applications of FT-IR spectroscopic methods in the study of the response of foodborne pathogenic bacteria to (sub)lethal environmental conditions.

A. Alvarez-Ordóñez and M. Prieto, *Fourier Transform Infrared Spectroscopy in Food Microbiology*, SpringerBriefs in Food, Health, and Nutrition, DOI 10.1007/978-1-4614-3813-7_3, © Avelino Alvarez-Ordóñez and Miguel Prieto 2012

3.1   Use of FT-IR Spectroscopy for the Assessment of the Mechanisms of Microbial Inactivation by Food Processing Technologies and Antimicrobial Compounds

The capacity of FT-IR spectroscopy as a tool to determine the cellular target of different antimicrobial compounds and food processing technologies has been shown in several studies (Feo et al. 2004; Schleicher et al. 2005; Bizani et al. 2005; Hu et al. 2007, 2009; Motta et al. 2008; Álvarez-Ordóñez and Prieto 2010).

Hu et al. (2009) combined FT-IR spectroscopy with spectral deconvolution and cluster analysis to investigate the biological effect (change of shape, function, biological physics, biochemistry, and composition of bacterial structures) of an ultrastrong static magnetic field on *Escherichia coli* and *Staphylococcus aureus*. The resulting cluster analysis indicated that the treatment provoked a significant effect on *E. coli* compared with *S. aureus*, with great changes in the spectral region from 1,500 to 1,200 cm^{-1}. The results obtained indicated that the secondary or tertiary helix structure of the DNA molecule was altered and that transcriptional increases or new gene fragments were generated in the DNA molecule. Furthermore, alterations in the structure of proteins and fatty acids were also found. The process of destruction of the cell wall and cell membrane of *E. coli* and *S. aureus* after treatment with AgI/TiO_2 photocatalyst under visible light irradiation was also monitored by using FT-IR measurement (Hu et al. 2007), and it was observed that the constituents of the cell envelopes disappeared and that some intermediates such as aldehydes, ketones, and carboxylic acids were formed.

Another application is the monitoring of the enzymatic repair of DNA, where some characteristic IR assignment bands can be assigned to repairing enzymes and the DNA substrate. Thus, the formation of thymidine dimers by ultraviolet light and also the light-driven DNA repair mechanisms by photolyase catalysis could be monitored in *E. coli* (Schleicher et al. 2005).

The effect of exposure to acid (pH 2.5), alkaline (pH 11.0), heat (55°C) and oxidative (40 mM H_2O_2) lethal conditions on the ultrastructure and global chemical composition of *Salmonella* Typhimurium using transmission electron microscopy and FT-IR spectroscopy combined with multivariate statistical methods (hierarchical cluster analysis and factor analysis) has been evaluated (Álvarez-Ordóñez and Prieto 2010). It could be concluded that all the treatment conditions caused important modifications in the five characteristic spectral regions (w_1, w_2, w_3, w_4, and w_5; see Sect. 1.1), although each lethal exposure showed specific effects on the different spectral ranges, with acid and alkaline stresses being responsible for the most relevant spectral modifications. Nevertheless, the most remarkable differences, even in the absence of any spectral transformation, were found in the w_4 spectral region (from 1,200 to 900 cm^{-1}), which suggests that some of the IR spectral changes could have been due to the presence of damage or conformational/compositional alterations in some or all of the components of the cell wall and cell membrane. The concordance of FT-IR data with ultrastructural changes observed by transmission electron microscopy corroborates the capacity of FT-IR spectroscopy as a tool to study the molecular transformations occurring in bacterial cells in response to environmental stress.

A number of studies have investigated the mode of action of bacteriocin-like substances by FT-IR spectroscopy. Bizani et al. (2005) showed that cerein 8A, a bacteriocin produced by the soil bacterium *Bacillus cereus* 8A, causes major changes in the spectral region corresponding to membrane fatty acids in *Listeria monocytogenes* and *B. cereus* and concluded that this antibacterial compound may cause dissipation of the proton-motive force as well as leakage of intracellular contents. Likewise, Motta et al. (2008) described the membrane effects on *L. monocytogenes* of BLS P34, a bacteriocin-like substance produced by the novel *Bacillus* sp. strain P34, and concluded that it acts through destabilization of the lipid packing or pore formation as some authors have proposed for other bacteriocins. Finally, FT-IR spectroscopy has been used to detect changes in the cellular composition of *Salmonella* Typhimurium cells after their treatment with antimicrobial compounds present in culture supernatants from *Lactobacillus fermentum* and *Lactobacillus plantarum* (Zoumpopoulou et al. 2010). Zoumpopoulou et al. (2010) detected by principal component analysis that the antimicrobial compounds produced by the lactobacilli interfered with the fatty acids of the cell membrane as well as the polysaccharides of the cell wall of *Salmonella* Typhimurium, pointing toward a dual killing mode.

FT-IR spectroscopy has also been successfully used to study the interactions of inorganic toxic substances with living bacterial cells (Feo et al. 2004). Feo et al. (2004) showed that the presence of selenium species at various concentrations in the culture medium exerted an effect on the IR spectra of *E. coli*, which predominantly differed in the spectral features characteristic of the outer-membrane components. Furthermore, this study proved the capacity of this technique not only to describe the interaction of selenium species with bacterial cells, but also to determine low concentrations of selenium species in aqueous samples.

3.2 FT-IR Spectroscopy as a Tool for Monitoring the Membrane Properties of Foodborne Microorganisms in Changing Environments

Most of the studies assessing the impact of processing technologies and suboptimal environmental conditions on the biochemical composition of bacterial cells by FT-IR spectroscopy have been focused on the examination of the membrane properties.

FT-IR spectroscopy was employed (Scherber et al. 2009) to quantify changes in the membrane phase behavior, as identified by the membrane phase transition temperature (T_m), of *E. coli* during desiccation and rehydration. For this purpose, the vibrational modes of the acyl chain –CH$_2$ symmetric stretching band at 2,850 cm^{-1} were monitored, and the peak location of the band was plotted as a function of temperature. T_m was determined by identifying the temperature at which the greatest change in the $-CH_2$ vibrational frequency occurred. Extensive cell desiccation (1 week at 20–40% relative humidity) resulted in an increase in T_m. Rehydration of *E. coli* resulted in a gradual regression of T_m, which began approximately 1 day after initial rehydration. Parallel examination of compositional changes in the bacterial membranes (determined by gas-chromatographic analysis of fatty acid methyl esters) showed that a correlation exists between the percent composition of saturated and unsaturated fatty acids and physiological changes of the membrane as determined by changes in membrane fluidity assessed by FT-IR analysis.

Also, an FT-IR spectroscopic approach has been used to assess the effects of osmotic pressure on the phase behavior of the polar lipid extracted from *E. coli* by monitoring the frequency shift of the 2,850-cm^{-1} band, assigned to the $-CH_2$ symmetric stretching mode, as a function of the osmotic pressure of the glycerol solution deposited over the phospholipid extract (Beney et al. 2004). The results obtained suggest that living cells respond to the increase in osmotic pressure by decreasing the membrane fluidity, that cell death occurs when membrane fluidity reaches a minimal value, and that dehydration promotes the transition from the Lα (liquid crystalline) to Lβ (gel) phase in the cytoplasmic membrane. These findings agree with those previously reported elsewhere (Mille et al. 2002), where a correlation was found between *E. coli* membrane modifications assessed by FT-IR spectroscopy and cell viability after combined osmotic and thermal treatments. Similar methods were used by other authors (Fang et al. 2007; Ami et al. 2009). Ami et al. (2009), studying lipid changes occurring in *E. coli* intact cells during recombinant protein expression and aggregation, described that protein misfolding and aggregation, commonly induced by exposure of bacteria to changing environmental conditions (e.g., heat and osmotic shock), induce rearrangements of the membrane and lipid composition in the host cells, with a reduction of membrane permeability and membrane fluidity and an increase of the $-CH_2$ infrared band intensity, which indicates the presence of longer and/or more saturated acyl chains under aggregation stress conditions. Finally, Fang et al. (2007) used FT-IR spectroscopy to investigate the effect of buckminsterfullerene ($_nC_{60}$), a fullerene-based nanomaterial, on the bacterial membrane lipid composition and phase behavior of *Pseudomonas putida* and *Bacillus subtilis*. This was the first demonstrated physiological adaptation response of bacteria to a manufactured nanomaterial, as shown by increased membrane fluidity for cells grown in the presence of high growth-inhibiting concentrations of $_nC_{60}$.

FT-IR spectroscopy has also been successfully used for monitoring the changes in *Salmonella* Typhimurium and *Salmonella* Enteritidis membrane fluidity as a function of growth temperature, growth medium pH, and NaCl concentration (Álvarez-Ordóñez et al. 2010). For cells grown in unsupplemented medium, an increase in growth temperature was linked to a decrease in membrane fluidity. Even though the effect of the NaCl concentration and the pH of the growth medium was considered of less importance, cells grown in acidified medium also showed a reduction in their membrane fluidity, and the supplementation of the culture medium with NaCl was associated with an increase in the bacterial membrane fluidity. Taking into account the well-known relationship between membrane fluidity and resistance to lethal treatments (Annous et al. 1999; Casadei et al. 2002; Alvarez-Ordoñez et al. 2008), the possibility to predict the bacterial membrane fluidity in changing environments commonly found in food is of great importance, since it may allow us to indirectly predict microbial resistance to several food processing agents and therefore the ability of microbes to cope with lethal challenges. In fact, FT-IR spectroscopy has occasionally been used to assess the relationship between the survival of foodborne pathogenic microorganisms when exposed to processing technologies and the structural changes of membrane phospholipids (Karatzas and Bennik 2002). Karatzas and Bennik (2002) tested an isolate of *L. monocytogenes* Scott A with high tolerance for high hydrostatic pressure and its wild-type counterpart and found that the highly resistant phenotype was not linked to the existence of an inherent difference in membrane fluidity between both strains.

FT-IR spectroscopy has also been used for the characterization and differentiation of mesophilic and thermophilic bacteria (with significantly higher protein-to-lipid ratio), by observing the change in the composition of the acyl chains (higher amount of saturated lipids in thermophilic bacteria) manifested in the absorption intensity of the CH(3) asymmetric stretching vibration (Garip et al. 2007).

3.3 Detection of Stress-Injured Microorganisms in Food-Related Environments by FT-IR Spectroscopy

FT-IR spectroscopic analysis has been applied to the qualitative and quantitative detection of injured pathogens in food products subjected to food preservation agents (Lin et al. 2004; Al-Qadiri et al. 2008a, b; Davis et al. 2010a). Al-Qadiri et al. (2008b) detected sublethally heat-injured and dead *Salmonella* Typhimurium and *L. monocytogenes* cells and discriminated between different heat treatment intensities. The variations in the spectral patterns of *Salmonella* Typhimurium and *L. monocytogenes* in the range from 1,800 to 1,300 cm^{-1} showed heat damage, with changes in the spectral properties of bacterial proteins, enzymes, and nucleic acids, with apparent denaturation. It also appeared likely that the spectral variations observed between 1,300 and 900 cm^{-1} were linked to damage of cell walls and cell membranes. These authors also predicted the degree of heat injury for both microorganisms, which indicates that FT-IR spectroscopy can also have applications for validating the effectiveness of various thermal processing treatments and for predicting the degree of cell injury or death. The same research group (Al-Qadiri et al. 2008a) studying in a mixed bacterial culture of *E. coli* and *Pseudomonas aeruginosa* the effect of chlorine-induced bacterial injury detected changes in the spectral features of bacterial ester functional groups of lipids, structural proteins, and nucleic acids, with apparent denaturation reflected between 1,800 and 1,300 cm^{-1} for injured bacterial cells, and suggested that FT-IR spectroscopy may be applicable for detecting the presence of injured and viable but not culturable waterborne pathogens that are underestimated or not discernible when using conventional microbial techniques.

The use of FT-IR spectroscopy combined with principal component analysis was successfully validated for the discrimination of intact and sonication-injured *L. monocytogenes* cells (Lin et al. 2004). Lin et al. (2004) detected changes in cells during sonication resulting from macromolecular shearing and subsequent redistribution of cell wall components with possible denaturation of intracellular proteins.

Recent studies have combined FT-IR spectroscopy with separation methods (filtration and immunomagnetic separation) and chemometrics (discriminant analysis and canonical variate analysis) to differentiate live and heat-treated bacterial cells in complex food matrixes. Live and heat-treated *Salmonella* Enteritidis and *Salmonella* Poona cells introduced onto chicken breast (Davis et al. 2010a) as well as live and dead *E. coli* O157:H7 cells introduced onto ground beef (Davis et al. 2010b) were discriminated and quantified.

3.4 Use of FT-IR Spectroscopy for the Study of Microbial Tolerance Responses

Several authors have proved the suitability of FT-IR spectroscopy to determine the global macromolecular changes occurring in bacteria in response to exposure to sublethal stressful environments and to study the plasticity of adaptive or tolerance responses in food-associated bacteria (Moen et al. 2005, 2009; Oust et al. 2006; Papadimitriou et al. 2008; Álvarez-Ordóñez et al. 2010).

Moen et al. (2005) and Oust et al. (2006) used FT-IR spectroscopy combined with multivariate analysis of variance and analysis of covariance (partial least squares regression), respectively, to study the changes in the phenotype of *Campylobacter jejuni* under nongrowth environmental conditions by investigating combinations of the factors temperature and oxygen tension, and reported an increase in the amount of polysaccharides and oligosaccharides under the nongrowth survival conditions probably linked to the production of a polysaccharide capsule important for the survival under harsh conditions. Similarly, Moen et al. (2009) assessed the biomolecular composition of an *E. coli* model strain exposed to diverse adverse conditions (sodium chloride, ethanol, glycerol, hydrochloric and acetic acids, sodium hydroxide, heat, cold, ethidium bromide, and the disinfectant benzalkonium chloride). These authors described an increase in the concentration of unsaturated fatty acids during exposure to ethanol and cold, and a decrease during exposure to acid and heat, with changes in the carbohydrate composition of the cells under refrigeration conditions.

Papadimitriou et al. (2008), who used FT-IR spectroscopy to study the acid tolerance response of *Streptococcus macedonicus*, segregated by principal component analysis of the second-derivative-transformed FT-IR spectra different acid-tolerant phenotypes in all spectral regions that are characteristic of major cellular constituents such as polysaccharides of the cell wall, fatty acids of the cell membrane, proteins, and other compounds that absorb in these regions. These findings provided evidence not only for the presence of significant alterations in major chemical constituents of *S. macedonicus* cells due to acid adaptation but also for a clear dependence of such alterations on the particular method employed to induce the tolerance response, and demonstrated the plasticity of the *S. macedonicus* acid tolerance response.

Likewise, Álvarez-Ordóñez et al. (2010) determined the effects of different adaptation conditions (temperature in the range 10–45°C, sodium chloride concentration in the range 0–4%, aerobic versus anaerobic growth, and acidification of the growth medium up to pH 4.5) on the FT-IR spectra of *Salmonella* Typhimurium and *Salmonella* Enteritidis. They showed that although all environmental factors tested affected the FT-IR spectra of *Salmonella* Typhimurium and *Salmonella* Enteritidis to some extent, growth temperature was the most influential factor, and the w_4 spectral region (from 1,200 to 900 cm⁻¹) was the most variable region, suggesting that *Salmonella* Typhimurium and *Salmonella* Enteritidis modulate their cell wall and cell membrane composition in response to shifts in growth temperature.

3.5 Applications of FT-IR Spectroscopy for the Study of Spore Composition

FT-IR spectroscopy seems to be an accurate analytical tool for monitoring the efficacy of sterilization techniques in inactivating spore-forming microorganisms. Various studies have shown the ability of FT-IR spectroscopy to evaluate spore properties in changing environments (Cheung et al. 1999; Perkins et al. 2004; Subramanian et al. 2006, 2007). Cheung et al. (1999) monitored the chemical changes of spore components of three *B. subtilis* strains by attenuated total reflectance FT-IR spectroscopy during spore germination. They observed modifications in the protein spectral region and release of calcium–dipicolinic acid during the germination process for the wild-type strain and for a deletion mutant in the *gerD* gene, encoding for the corresponding germinant receptor GerD, whereas these spectral changes were not apparent for a *gerA* deletion mutant, which is unable to synthesize the GerA germinant receptor and which was not able to successfully complete the germination process. Several authors (Perkins et al. 2004; Subramanian et al. 2006, 2007) have used an FT-IR spectroscopic approach to study some of the biochemical changes occurring in bacterial spores during food processing (i.e., autoclaving, thermal processing, and pressure-assisted thermal processing). Perkins et al. (2004) evaluated the effect of the autoclaving process on bacterial endospores of *B. cereus, Bacillus atrophaeus*, *Bacillus megaterium*, *B. subtilis*, and *Clostridium perfringens* by reflectance FT-IR microspectroscopy, and they found considerable changes occurring in the amide I and II regions in samples that had undergone autoclaving as a result of secondary structure conversions to intermolecular β-sheets, due to protein denaturing and subsequent aggregation during the autoclaving, as well as the loss of the absorption band at 1,750 cm⁻¹, assigned to dipicolinic acid, indicating release of this compound during the autoclaving process. These biochemical modifications allowed them to successfully discriminate between intact and autoclaved spores by means of a principal component analysis, and to use dipicolinic acid as a marker of spore germination or inactivation in *Bacillus* spp. (Goodacre et al. 2000; Perkins et al. 2005). Similarly, Subramanian et al. (2007) showed for *Clostridium tyrobutyricum*, *Bacillus sphaericus*, and three strains of *Bacillus amyloliquefaciens* that pressure-assisted thermal processing caused a change in α-helices and β-sheets of secondary proteins, which was evident in the spectral regions from 1,655 to 1,626 cm^{-1}, and a decrease in the intensity of absorption bands from dipicolinic acid (1,281, 1,378, 1,440, and 1,568 cm^{-1}), indicating release of this compound during the initial stages of the pressure-assisted thermal treatment. On the other hand, although thermal processing alone also rendered changes in the bands associated with secondary proteins, little or no modification was observed in dipicolinic acid bands. In addition, this study found a correlation between the spore content of dipicolinic acid, as assessed by FT-IR spectroscopy, and the resistance of spores to pressure-assisted thermal processing, a finding which highlights the potential use of this technique for rapid screening of spores that are highly resistant to pressure-assisted thermal processing. This research group (Subramanian et al. 2006) had previously demonstrated for the same bacterial strains the potential of FT-IR spectroscopy combined with multivariate regression analysis to predict on the basis of differences in biochemical composition viable spore concentrations in samples treated by pressure-assisted thermal processing and thermal processing.

Chapter 4
Fourier Transform Infrared Spectroscopic Methods for Microbial Ecology

Fourier transform (FT-IR) spectroscopy can be utilized to detect differences in microbial community structures, including binary mixed cultures of two microorganisms, where it is used to distinguish and quantify microbial populations.

4.1 Assessment of Dynamic Changes in Microbial Populations by FT-IR Spectroscopy

A novel potential application of FT-IR spectroscopy is the study of dynamic changes in bacterial populations.

Becker et al. (2006) used this technique as a rapid tool for the examination of different subpopulations of *Staphylococcus aureus*. They discriminated between small-colony variants, recognized as the causative organisms in chronic, recurrent, and antibiotic-resistant infections, and the normal phenotypes of this microorganism, and investigated dynamic processes of reversion of small-colony variants to the normal phenotype and vice versa using first-derivative IR spectra combined with hierarchical cluster analysis. Similarly, the technique was used to differentiate wild-type strains of *Listeria monocytogenes* from their spontaneous sakacin P resistant mutants obtained after single exposure to sakacin P, a class IIa bacteriocin that is active against foodborne pathogens (Tessema et al. 2009). Tessema et al. (2009) showed that resistance to sakacin P introduced distinct changes mainly in the regions corresponding to polysaccharides, fatty acids, and proteins, which represent the major cell membrane components involved in the resistant phenotype developed.

Several studies have investigated the biological heterogeneity of microbial growth using an FT-IR spectroscopic approach (Choo-Smith et al. 2001; Filip et al. 2004; Ede et al. 2004; Ngo Thi and Naumann 2007). Choo-Smith et al. (2001) studied the time and spatial heterogeneity in the development of microcolonies of *Escherichia coli*, *S. aureus*, and *Candida albicans* by means of Raman and FT-IR microspectroscopy. It was found that whereas there was little spectral variance in 6-h microcolonies, significant

A. Alvarez-Ordóñez and M. Prieto, *Fourier Transform Infrared Spectroscopy in Food Microbiology*, SpringerBriefs in Food, Health, and Nutrition, DOI 10.1007/978-1-4614-3813-7_4, © Avelino Alvarez-Ordóñez and Miguel Prieto 2012

colony heterogeneity occurred in the strains cultured for 12 and 24 h, which was attributed to higher glycogen content in the surface layers and to increased levels of carotenoid pigmentation in certain *S. aureus* strains in addition to a relatively higher RNA content in the deeper layers of the colony. These findings suggest that the colony is composed of older cells in the surface layers and younger, actively dividing cells in the deeper layers. Similar observations were made by Ngo Thi and Naumann (2007), who studying macrocolonies of *Legionella bozemanii, Bacillus megaterium,* and *C. albicans* with an FT-IR microspectroscopic mapping approach described that the levels of the storage material poly(β-hydroxybutyric acid) in *L. bozemanii* and levels of the capsular compound in *B. megaterium* were significantly higher in cells at the surface of the colonies than in those growing at the bottom next to the agar surface. Furthermore, significant changes in the lipid, protein, and carbohydrate composition of *C. albicans* colonies were found depending on where within the colonies the cells grew. The promising results obtained in these latter studies suggest that FT-IR spectroscopy can be extended to the study of other complex microbial communities, such as biofilms. In fact, recent studies (Comeau et al. 2009; Holman et al. 2009) have already developed an FT-IR spectromicroscopy approach to directly monitor and map cellular changes in *E. coli* and *Pseudomonas aeruginosa* biofilms.

Other studies have used FT-IR spectroscopy to monitor structural changes throughout the growth process (Filip et al. 2004; Ede et al. 2004; Castro et al. 2010). Ede et al. (2004) investigated structural changes occurring in the cells of *Bacillus stearothermophilus, Halobacterium salinarum, Halococcus morrhuae,* and *Acetobacter aceti* during growth by FT-IR spectroscopy using the attenuated total reflectance (ATR) sampling technique. For all species they found significant spectral changes, indicating structural changes in the cells during increases in cell numbers. The major change for *B. stearothermophilus* was in the lipid content, which was maximum during the exponential phase of the growth curve. For the halophiles *H. salinarum* and *H. morrhuae,* the major change was that the concentration of sulfate ion in the cells varied during the growth curve and was higher during the middle of the exponential phase of the growth curve. *A. aceti* cells showed increasing polysaccharide content during the growth curve as well as maximum lipid content during the exponential phase of growth. Filip et al. (2004) demonstrated for *B. subtilis* the existence of a relevant variability in FT-IR spectra as a function of the culture age and the quality of the nutrient growth solution used. Finally, Castro et al. (2010) used FT-IR spectroscopy to assess the effects of oxygen tension during bacterial growth on the cell surface properties and biochemical composition of the pathogens *E. coli* O157:H7 and *Yersinia enterocolitica.*

4.2 Use of FT-IR Spectroscopy to Identify and Quantify Microorganisms in Binary Mixed Cultures

FT-IR spectroscopy has shown capacity to distinguish and quantify microbial populations in binary mixed cultures of two microorganisms. Quantitative differentiation of individual species present in a mixed population is achieved by determining the ratios of different microorganisms in the mixture. This application has been assayed in binary mixed cultures of *E. coli* O157:H7 and *Alicyclobacillus* spp. (Al-Qadiri et al. 2006), *Pseudomonas putida* and *Rhodococcus ruber* (Schawe et al. 2011), yeasts (*Saccharomyces cerevisiae/Hanseniaspora uvarum*, Oberreuter et al. 2000; *Saccharomyces cerevisiae/Debaryomyces hansenii/Rhodotorula minuta*, Rellini et al. 2009), and diverse bacteria (*Staphylococcus aureus* and *Lactococcus lactis*, Nicolaou et al. 2011; *Lactobacillus acidophilus/Streptococcus salivarius* ssp. *Thermophilus*, Oberreuter et al. 2000). With use of FT-IR microspectroscopy, identification of mixed microorganisms grown on Petri dishes directly plated from dilutions of mixed community was also successfully achieved (Wenning et al. 2006).

4.3   Use of FT-IR Spectroscopy to Detect Food Spoilage due to Microbial Activity

Biochemical changes within the food substrate due to microbial activity can be monitored using FT-IR spectroscopy and serve as a tool to detect food spoilage. This application was first successfully demonstrated for chicken meat (Ellis et al. 2002), where a correlation between metabolic profiles and microbial counts was obtained. Ellis et al. (2002) used ATR FT-IR spectroscopy and partial least squares regression, which made a quantitative interpretation of FT-IR spectra possible. Other authors have proposed similar methods to detect spoilage in minced beef (Ammor et al. 2008), minced pork (Papadopoulou et al. 2011), and pasteurized milk (Nicolaou and Goodacre 2008). According to the authors, the FT-IR spectrum may be considered as a metabolic fingerprint and reflects the microbial and chemical condition of fresh foods.

Chapter 5
Conclusions and Future Prospects

The advantages of Fourier transform infrared (FT-IR) spectroscopy and the particular applications in food microbiology are now well recognized, and the number of research groups working with this method is steadily expanding. The capacity of FT-IR spectroscopy to taxonomically characterize bacteria is well established since a large number of bacterial genera and species have been successfully identified using it. The infrared (IR) spectrum of microbial biomass represents a reflection of the overall chemical composition, which is valuable in taxonomy studies. FT-IR spectroscopy can afford information additional to phenotypic and genotypic data which may help to establish a more robust taxonomic classification. There is evidence that some compounds and structures (mainly the cell membrane and cell wall) appear to have a stronger influence on the IR spectrum. For identification and typing, good identification results depend on the quality and size of the databases as well as on an adequate mathematical procedure. Standardization of conditions is essential to obtain good-quality IR spectra in order to build up reliable databases to be able to assign unknowns and to detect outliers. Foodborne bacteria constantly face fluctuations in environmental conditions which result in (sub)lethal stress exposures. Consequently, they have evolved adaptive networks to cope with the challenges of a changing environment and to survive stress conditions. A better understanding of the mechanisms of bacterial inactivation by food processing technologies and the molecular changes occurring in the cell as part of the bacterial stress response would therefore lead to the optimization of current food processing strategies. The recent availability of numerous genome sequences has catalyzed a burst of genomics-driven fundamental research in the ecology, physiology, and virulence of foodborne pathogens. In addition, the development of novel methods has led to an increasing number of transcriptomics and proteomics studies from which changes in the chemical composition of cells in response to stress conditions and functions of different genes and macromolecules have been predicted. However, only some of these predictions have actually been verified directly. This review shows that FT-IR spectroscopy, which is an adequate tool to understand how environmental conditions affect the whole cell, could be successfully combined with

A. Alvarez-Ordóñez and M. Prieto, *Fourier Transform Infrared Spectroscopy*
in Food Microbiology, SpringerBriefs in Food, Health, and Nutrition,
DOI 10.1007/978-1-4614-3813-7_5, © Avelino Alvarez-Ordóñez and Miguel Prieto 2012

novel transcriptomics and statistical methods in order to find a new hypothesis for the survival mechanisms of foodborne pathogenic bacteria (Table 5.1).

The detection and identification of pathogens in foods is a basic cornerstone of food safety because it makes it possible to identify sources of contamination, provides data on the evaluation of risk-reduction measures, and identifies the food chain operations, processes, batches, or products representing a threat to public health. Furthermore, it is also fundamental in the epidemiological investigation of foodborne diseases. The presence of stress-injured bacterial cells in foods represents a challenge to those involved in food quality assurance, as routine microbiological procedures may yield negative results for sublethally injured cells. Thus, food could be presumed to be safe and free from pathogenic cells but during storage could become dangerous owing to recovery and growth of previously injured cells. The existence of phenotypic heterogeneity within bacterial populations also represents a threat to food safety through the expression of highly resistant phenotypes under nonoptimal environmental conditions. Therefore, the capacity recently shown by FT-IR spectroscopy for the detection and discrimination of stress-injured bacteria and phenotypic variants of a given bacterial population shows promise for this spectroscopic technique as a tool for the analysis of the microbiological quality of foods.

During the past few years, biomarkers for bacterial resistance have emerged as an indispensable tool for industrial companies, as they can be used to predict the condition of the cell and to understand the mechanisms of stress adaptation. This review has highlighted the usefulness of FT-IR spectroscopy for monitoring two potential biomarkers of bacterial stress resistance, i.e., the membrane fluidity and the spore content of dipicolinic acid. Taking into account that microbial resistance to several food processing technologies is linked to membrane and spore properties, the possibility to predict membrane fluidity and spore content of dipicolinic acid by FR-IR spectroscopy in changing environments commonly found in the food chain would allow an indirect prediction of microbial resistance to food preservation agents. Future work is needed using combined FT-IR and predictive microbiology approaches in order to develop and validate effective predictive models.

In natural and industrial settings, bacteria usually live in consortia with other bacterial species in structures named biofilms. Although biofilms are a constant concern in food processing environments, many essential features of biofilm cells remain unclear. Recent reports describing the ability of FT-IR spectroscopy to study the spatial and temporal dynamics of complex bacterial communities show the potential of this vibrational technique to determine the physiological changes occurring in both single-species and multispecies biofilms.

Nevertheless, despite the great potential of FT-IR spectroscopy for performing microbial ecology studies, it should be taken into account that the compositional and structural variations determined by FT-IR spectroscopy are, in general, very complex. Therefore, acquisition of good-quality spectra and combination of the technique with accurate mathematical deconvolution procedures is essential. Furthermore, it seems clear that for assignment to distinct cellular components, complementary data derived from independent techniques (i.e., electron microscopy, scanning confocal laser microscopy, biochemical analysis or DNA microarray analysis) are necessary, as no single cellular or molecular characterization method embraces all the desirable properties.

Table 5.1 Scientific articles published showing applications of Fourier transform infrared (*FT-IR*) spectroscopy in food microbiology (taxonomy and identification of foodborne bacteria, mechanisms of bacterial inactivation, stress response and molecular changes occurring in foodborne bacteria under different conditions, food spoilage and food ecology)

Aim	Microorganism	Type of mathematical analysis	FT-IR technique	References
Taxonomy and identification of foodborne bacteria				
Differentiation at subspecies level	*Campylobacter jejuni* and *Campylobacter coli*	HCA, ANN	Transmittance (ZnSe windows)	Mouwen et al. (2005, 2006)
Differentiation at subspecies level	*Yersinia enterocolitica*	ANN	Transmittance (ZnSe windows)	Kuhm et al. (2009)
Classification at species and subspecies level	*Salmonella*	HCA	Transmittance	Seltmann et al. (1994)
Differentiation at subspecies level	*Salmonella enterica*	CVA, DA	Reflectance microspectroscopy	Kim et al. (2005, 2006)
Differentiation at subspecies level	*Salmonella enterica*	SIMCA	Transmittance (ZnSe), ATR	Baldauf et al. (2006)
Differentiation at subspecies level	*Salmonella enterica* Enteritidis	PLS-DA	Transmittance (silicon plate)	Preisner et al. (2010)
Differentiation at subspecies level	*Escherichia coli*	PCA, CVA	Reflectance, microspectroscopy	Kim et al. (2006)
	Bacillus cereus			Lin et al. (1998)
Differentiation at species and subspecies level	*Bacillus cereus* and other *Bacillus* species	HCA	Transmittance (ZnSe)	Mietke et al. (2010)
Differentiation at species and subspecies level	*Brucella* spp.	LDA	–	Miguel Gomez et al. (2003)
Differentiation at species and subspecies level	*Enterobacter (Cronobacter) sakazakii*	PCA, SIMCA	–	Lin et al. (2010)
Differentiation at species level	*Listeria* spp.	PCA, CVA	Transmittance	Holt et al. (1995)
Differentiation at species and subspecies level	*Listeria monocytogenes*	CVA	Transmittance	Lefier et al. (1997)

(continued)

Table 5.1 (continued)

Aim	Microorganism	Type of mathematical analysis	FT-IR technique	References
Differentiation at subspecies level	*Listeria monocytogenes*	HCA, ANN	Transmittance	Rebuffo-Scheer et al. (2007)
Differentiation at subspecies level	*Listeria monocytogenes*	DA, PLSR	Microspectroscopy, transmittance	Janbu et al. (2008)
Differentiation at subspecies level	*Listeria monocytogenes*	CVA, HCA	Microspectroscopy, ATR	Davis and Mauer (2011)
Differentiation at subspecies level	Group I and group II *Clostridium botulinum*	HCA	Focal plane array	Kirkwood et al. (2006)
Differentiation of groups	Coagulase-negative and coagulase-positive staphylococci	PCA, HCA, ANN	Transmittance (ZnSe windows)	Amiali et al. (2007, 2008)
Differentiation at subspecies level	*Mycobacterium bovis*	PCA, DFA, HCA	Reflectance and transmittance	Winder et al. (2006)
Differentiation at species and subspecies level	LAB from plants, milk and cheese	DA, PCA	Transmittance (ZnSe windows)	Weinrichter et al. (2001)
Differentiation at species and subspecies level	LAB from cheese	DA	Transmittance	Amiel et al. (2000)
Differentiation at species and subspecies level	LAB from meat, fish and cheese	HCA, PLSR, SIMCA, KNN	Transmittance (ZnSe windows)	Oust et al. (2004)
Differentiation at species level	LAB from cheese	HCA, PCA	Transmittance	Savic et al. (2008)
Identification and differentiation at species level	LAB from meat products	HCA, SDA	Transmittance (ZnSe windows)	Mouwen et al. (2011)
Differentiation at species level	Heterofermentative and homofermentative LAB from kefir	HCA	–	Bosch et al. (2006)
Differentiation at species level	*Bacillus* spp.	CVA	Transmittance	Beattie et al. (1998)
Differentiation at species level	*Bacillus* spp.	PCA, HCA	ATR	Brandes-Ammann and Brandl (2011)

Description	Organism	Chemometric method	Spectroscopy mode	Reference
Differentiation at species and subspecies level	*Bacillus, Escherichia coli, Salmonella, Listeria*	PCA, SIMCA	ATR	Al-Holy et al. (2006)
Differentiation at species level	*Staphylococcus aureus* and other *Staphylococcus* species	PCA, DA, CVA	Transmittance	Lamprell et al. (2006)
Classification and identification at species level	Yeasts	HCA	Transmittance	Kummerle et al. (1998)
Classification at species level	Yeasts	HCA	Transmittance (ZnSe windows)	Paramithiotis et al. (2000)
Identification at species level	Yeasts	HCA	Microspectroscopy	Wenning et al. (2002)
Classification and identification at genus, species and subspecies level	Filamentous fungi (*Aspergillus* and *Penicillium*)	HCA	Transmittance (ZnSe windows)	Fischer et al. (2006)
Assessment of the mechanisms of bacterial inactivation by food processing technologies and antimicrobial compounds				
Study the mechanisms of inactivation by heat, acid, alkaline and oxidative stress	*Salmonella* Typhimurium	HCA, FA	Transmittance (ZnSe windows)	Álvarez-Ordóñez and Prieto (2010)
Study the interaction of selenium with bacterial cells	*Escherichia coli*	HCA, PCA	Transmittance (ZnSe windows)	Feo et al. (2004)
Investigate the mode of action of the bacteriocin cerein 8A	*Listeria monocytogenes, Bacillus cereus*	–	ATR	Bizani et al. (2005)
Investigate the mode of action of the bacteriocin BLS P34	*Listeria monocytogenes*	–	ATR	Motta et al. (2008)
Monitor the destruction of the cells wall by photocatalytic degradation with AgI/TiO$_2$ under visible light irradiation	*Escherichia coli, Staphylococcus aureus*	–	Transmittance (KBr pellets)	Hu et al. (2007)
Investigate the biological effect of ultrastrong static magnetic fields	*Escherichia coli*	HCA, deconvolution	Transmittance (KBr pellets)	Hu et al. (2009)
Detect changes in cellular composition in response to antimicrobial compound(s) from *Lactobacillus* strains	*Salmonella* Typhimurium	PCA	Transmittance (ZnSe windows)	Zoumpopoulou et al. (2010)

(continued)

Table 5.1 (continued)

Aim	Microorganism	Type of mathematical analysis	FT-IR technique	References
Detect formation of tymidine dimers by UV light and study light-driven DNA repair mechanisms	*Escherichia coli*	—	(Difference) transmittance	Schleicher et al. (2005)
Monitoring of membrane properties in changing environments				
Study the membrane phase behavior during desiccation, rehydration and growth recovery	*Escherichia coli*	t test for the location of the CH_2 symmetric stretching band	Transmittance (CaF_2 windows)	Scherber et al. (2009)
Monitor changes in membrane structure after osmotic stress	*Escherichia coli*	—	Transmittance (ZnSe windows)	Beney et al. (2004)
Study the influence of temperature and osmotic pressure on the membrane's physical structure	*Escherichia coli*	—	Transmittance (ZnSe windows)	Mille et al. (2002)
Investigate the effects of recombinant protein misfolding and aggregation on bacterial membranes	*Escherichia coli*	—	Microspectroscopy (BaF_2 windows)	Ami et al. (2009)
Investigate the effect of a fullerene-based nanomaterial on bacterial membrane phase behavior	*Bacillus subtilis, Pseudomonas putida*	—	Microspectroscopy (aluminum pan)	Fang et al. (2007)
Investigate the relationship between membrane fluidity and resistance to high hydrostatic pressure	*Listeria monocytogenes*	—	Microspectroscopy	Karatzas and Bennik (2002)
Detection and differentiation of stress-injured microorganisms in food-related environments				
Detect the presence of chlorine-injured bacteria	*Escherichia coli, Pseudomonas aeruginosa*	PCA, SIMCA	ATR	Al-Qadiri et al. (2008a)
Detection of sublethal thermal injury. Discriminate between heat-injured and live bacteria	*Salmonella* Typhimurium, *Listeria monocytogenes*	SIMCA, PLSR	ATR	Al-Qadiri et al. (2008b)

Description	Organism	Analysis	Technique	Reference
Discrimination between intact and sonication-injured cells	*Listeria monocytogenes*	PCA	ATR	Lin et al. (2004a)
Differentiation of live and heat-treated cells in chicken breast	*Salmonella* Typhimurium	DA, CVA, PCA, PLSR	Microspectroscopy	Davis et al. (2010a)
Differentiation of live and dead cells in ground beef	*E. coli* O157:H7	DA, CVA, PLSR	Microspectroscopy	Davis et al. (2010b)
Assessment of dynamic changes in bacterial populations				
Study of the dynamic changes in *S. aureus* small-colony variants	*Staphylococcus aureus*	HCA	NA	Becker et al. (2006)
Detect bacterial variants resistant to sakacin P	*Listeria monocytogenes*	PCA	NA	Tessema et al. (2009)
Investigate microbial (micro) colony heterogeneity	*Escherichia coli, Staphylococcus aureus, Candida albicans*	HCA	Microspectroscopy	Choo-Smith et al. (2001)
Investigate the heterogeneity of cell growth in microbial colonies	*Legionella bozemanii, Bacillus megaterium, Candida albicans*	HCA	Microspectroscopy	Ngo Thi and Naumann (2007)
Monitor bacterial activity in biofilms	*Escherichia coli*	—	Microspectroscopy	Holman et al. (2009)
Investigate the growth of biofilms on ZnSe and TiO$_2$-coated ZnSe internal reflection elements	*Pseudomonas aeruginosa*	—	ATR	Comeau et al. (2009)
Assessment of structural changes in bacterial cells during population growth	*Bacillus stearothermophilus, Halobacterium salinarum, Halococcus morrhuae, Acetobacter aceti*	—	ATR	Ede et al. (2004)
Study the influence of culture age and quality of the nutrient growth solution on FT-IR spectra	*Bacillus subtilis*	—	Transmittance (KBr pellets), ATR	Filip et al. (2004)
Investigate the effects of dissolved oxygen tension during bacterial growth and acclimation on the cell surface properties and biochemical composition	*Escherichia coli, Yersinia enterocolitica*	—	ATR	Castro et al. (2010)

(continued)

Table 5.1 (continued)

Aim	Microorganism	Type of mathematical analysis	FT-IR technique	References
Study of bacterial tolerance responses				
Study the changes in the phenotype under non-growth environmental conditions (factors temperature and oxygen tension)	*Campylobacter jejuni*	MANOVA	Transmittance (ZnSe windows)	Moen et al. (2005)
Study the changes in the phenotype under non-growth environmental conditions (factors temperature and oxygen tension)	*Campylobacter jejuni*	PLSR	Transmittance (ZnSe windows)	Oust et al. (2006)
Assess the biomolecular composition after exposure to sodium chloride, ethanol, glycerol, hydrochloric and acetic acids, sodium hydroxide, heat, cold, ethidium bromide and benzalkonium chloride	*Escherichia coli*	PCA	Transmittance (ZnSe windows)	Moen et al. (2009)
Study the acid tolerance response	*Streptococcus macedonicus*	PCA	Transmittance (ZnSe windows)	Papadimitriou et al. (2008)
Determine molecular changes in changing environments of temperature, pH and salt concentration	*Salmonella* Typhimurium, *Salmonella* Enteritidis	HCA	Transmittance (ZnSe windows)	Álvarez-Ordóñez et al. (2010)
Study of spore ecology				
Monitor chemical changes during spore germination	*Bacillus subtilis*	–	ATR	Cheung et al. (1999)
Evaluate the effect of the autoclaving process on bacterial endospores	*Bacillus cereus, Bacillus atrophaeus, Bacillus megaterium, Bacillus subtilis, Clostridium perfringens*	PCA	Microspectroscopy	Perkins et al. (2004)

			Microspectroscopy	Perkins et al. (2005)
Use of dipicolinic acid as a marker of spore germination or inactivation	*Bacillus subtilis*	–	Microspectroscopy	Perkins et al. (2005)
Use of dipicolinic acid as a marker of spore germination or inactivation	*Bacillus amyloliquefaciens, Bacillus cereus, Bacillus licheniformis, Bacillus megaterium, Bacillus subtilis* (including *B. niger* and *B. atrophaeus*), *Bacillus sphaericus*, and *Brevi laterosporus*	Rule induction and genetic programming	Diffuse reflectance absorbance	Goodacre et al. (2000)
Determination of spore inactivation during thermal and pressure-assisted thermal processing	*Bacillus amyloliquefaciens, Bacillus subtilis sphaericus, Clostridium tyrobutyricum*	SIMCA, PLSR	ATR	Subramanian et al. (2006)
Monitor biochemical changes in bacterial spore during thermal and pressure-assisted thermal processing	*Bacillus amyloliquefaciens, Bacillus sphaericus, Clostridium tyrobutyricum*	SIMCA	ATR	Subramanian et al. (2007)
Food spoilage				
Monitor biochemical changes in poultry before spoilage	Poultry	PLSR	ATR	Ellis et al. (2002)
Monitor biochemical changes in pasteurized milk before spoilage	Pasteurized milk	PCA-DA and PLSR	ATR	Nicolaou and Goodacre (2008)
Monitor biochemical changes in minced beef before spoilage	Minced beef	PCA-DA and PLSR	ATR	Ammor et al. (2008)
Monitor biochemical changes in minced pork before spoilage	Minced pork	PCA-DA and PLSR	ATR	Papadopoulou et al. (2011)
Identification and quantification of bacteria in binary mixed cultures				
Escherichia coli O157:H7/*Alicyclobacillus* spp.		PCA, SIMCA	–	Al-Qadiri et al. (2006)
Escherichia coli/*Pseudomonas aeruginosa*		PCA, SIMCA	–	Al-Qadiri et al. (2006)

(continued)

Table 5.1 (continued)

Aim	Microorganism	Type of mathematical analysis	FT-IR technique	References
Pseudomonas putida/*Rhodococcus ruber*		PCA	Transmittance (ZnSe windows)	Schawe et al. (2011)
Saccharomyces cerevisiae/*Hanseniaspora uvarum*		PLSR	Transmittance (ZnSe windows)	Oberreuter et al. (2000)
Saccharomyces cerevisiae/*Debaryomyces hansenii*/*Rhodotorula minuta*		PcoA	Transmittance (silicon windows)	Rellini et al. (2009)
Lactobacillus acidophilus/*Streptococcus salivarius* ssp. *thermophilus*		PLSR	Transmittance (ZnSe windows)	Oberreuter et al. (2000)
Staphylococcus aureus and *Lactococcus lactis* ssp.		PLS		Nicolaou et al. (2011)

ANN artificial neural network, *ATR* attenuated total reflectance, *CVA* canonical variate analysis, *DA* discriminant analysis, *FA* factor analysis, *HCA* hierarchical cluster analysis, *KNN* k-nearest neighbors, *LAB* lactic acid bacteria, *LDA* linear discriminant analysis, *MANOVA* multivariate analysis of variance, *NA* not available, *PCA* principal component analysis, *PcoA* principal coordinate analysis, *PLS-DA* partial least squares–discriminant analysis, *PLSR* partial least squares regression, *SDA* stepwise discriminant analysis, *SIMCA* soft independent modeling of class analogy

Acknowledgments

The support of the Spanish Ministry of Science and Innovation (Plan Nacional I+D+I, project reference AGL2008–02668) and Junta de Castilla y León (project reference LE049A05) is gratefully acknowledged.

A. Alvarez-Ordóñez and M. Prieto, *Fourier Transform Infrared Spectroscopy in Food Microbiology*, SpringerBriefs in Food, Health, and Nutrition, DOI 10.1007/978-1-4614-3813-7, © Avelino Alvarez-Ordóñez and Miguel Prieto 2012

Abbreviations

ATR   Attenuated total reflectance
DRIFT Diffuse reflectance Fourier transform infrared
DTGS   Deuterated triglycine sulfate
FT-IR Fourier transform infrared
IR   Infrared
LPS   Lipopolysaccharide
MCT   Mercury cadmium telluride
MIR   Mid infrared

A. Alvarez-Ordóñez and M. Prieto, *Fourier Transform Infrared Spectroscopy* 47
in Food Microbiology, SpringerBriefs in Food, Health, and Nutrition,
DOI 10.1007/978-1-4614-3813-7, © Avelino Alvarez-Ordóñez and Miguel Prieto 2012

References

Al-Holy, M., Lin, M., Al-Qadiri, H., Cavinato, A.G., Rasco, B.A., 2006. Classification of foodborne pathogens by Fourier transform infrared spectroscopy and pattern recognition techniques. J. Rapid Methods Autom. Microbiol. 14, 189–200.

Al-Qadiri, H.M., Al-Alami, N.I., Al-Holy, M.A., Rasco, B.A., 2008a. Using Fourier transform infrared (FT-IR) absorbance spectroscopy and multivariate analysis to study the effect of chlorine-induced bacterial injury in water. J. Agric. Food Chem. 56, 8992–8997.

Al-Qadiri, H.M., Lin, M., Al-Holy, M.A., Cavinato, A.G., Rasco, B.A., 2008b. Detection of sublethal thermal injury in *Salmonella enterica* serotype typhimurium and *Listeria monocytogenes* using Fourier transform infrared (FT-IR) spectroscopy (4000 to 600 cm(-1)). J. Food Sci. 73, M54–M61.

Al-Qadiri, H.M., Lin, M., Cavinato, A.G., Rasco, B.A., 2006. Fourier transform infrared spectroscopy, detection and identification of *Escherichia coli* O157:H7 and *Alicyclobacillus strains* in apple juice. Int. J. Food Microbiol. 111, 73–80.

Alvarez-Ordóñez, A., Fernandez, A., Lopez, M., Arenas, R., Bernardo, A., 2008. Modifications in membrane fatty acid composition of *Salmonella* typhimurium in response to growth conditions and their effect on heat resistance. Int. J. Food Microbiol. 123, 212–219.

Álvarez-Ordóñez, A., Halisch, J., Prieto, M., 2010. Changes in Fourier transform infrared spectra of *Salmonella enterica* serovars Typhimurium and Enteritidis after adaptation to stressful growth conditions. Int. J. Food Microbiol. 142, 97–105.

Álvarez-Ordóñez, A., Prieto, M., 2010. Changes in ultrastructure and Fourier transform infrared spectrum of *Salmonella enterica* serovar Typhimurium cells after exposure to stress conditions. Appl. Environ. Microbiol. doi:10.1128/AEM.00312–10.

Ami, D., Natalello, A., Schultz, T., Gatti-Lafranconi, P., Lotti, M., Doglia, S.M., de, M.A., 2009. Effects of recombinant protein misfolding and aggregation on bacterial membranes. Biochim. Biophys. Acta 1794, 263–269.

Amiali, N.M., Mulvey, M.R., Berger-Bachi, B., Sedman, J., Simor, A.E., Ismail, A.A., 2008. Evaluation of Fourier transform infrared spectroscopy for the rapid identification of glycopeptide-intermediate Staphylococcus aureus. J. Antimicrob. Chemother. 61, 95–102.

Amiali, N.M., Mulvey, M.R., Sedman, J., Louie, M., Simor, A.E., Ismail, A.A., 2007. Rapid identification of coagulase-negative staphylococci by Fourier transform infrared spectroscopy. J. Microbiol. Methods 68, 236–242.

Amiel, C., Mariey, L., Curk-Daubie, M.C., Pichon, P., Travert, J., 2000. Potentiality of Fourier Transform Infrared Spectroscopy (FTIR) for discrimination and identification of dairy Lactic acid bacteria. Lait 80, 445–459.

Ammor, M.S., Argyri, A., Nychas, G.J., 2008. Rapid monitoring of the spoilage of minced beef stored under conventionally and active packaging conditions using Fourier transform infrared spectroscopy in tandem with chemometrics. Meat Sci.

A. Alvarez-Ordóñez and M. Prieto, *Fourier Transform Infrared Spectroscopy in Food Microbiology*, SpringerBriefs in Food, Health, and Nutrition, DOI 10.1007/978-1-4614-3813-7, © Avelino Alvarez-Ordóñez and Miguel Prieto 2012

Annous, B.A., Kozempel, M.F., Kurantz, M.J., 1999. Changes in membrane fatty acid composition of *Pediococcus* sp. strain NRRL B-2354 in response to growth conditions and its effect on thermal resistance. Appl. Environ. Microbiol. 65, 2857–2862.

Baldauf, N.A., Rodriguez-Romo, L.A., Yousef, A.E., Rodriguez-Saona, L.E., 2006. Differentiation of selected *Salmonella enterica* serovars by Fourier transform mid-infrared spectroscopy. Appl. Spectrosc. 60, 592–598.

Beattie, S.H., Holt, C., Hirst, D., Williams, A.G., 1998. Discrimination among *Bacillus cereus, B. mycoides* and *B. thuringiensis* and some other species of the genus *Bacillus* by Fourier transform infrared spectroscopy. FEMS Microbiol. Lett. 164, 201–206.

Becker, K., Laham, N.A., Fegeler, W., Proctor, R.A., Peters, G., von, E.C., 2006. Fourier-transform infrared spectroscopic analysis is a powerful tool for studying the dynamic changes in *Staphylococcus aureus* small-colony variants. J. Clin. Microbiol. 44, 3274–3278.

Beney, L., Mille, Y., Gervais, P., 2004. Death of *Escherichia coli* during rapid and severe dehydration is related to lipid phase transition. Appl. Microbiol. Biotechnol. 65, 457–464.

Bizani, D., Motta, A.S., Morrissy, J.A., Terra, R.M., Souto, A.A., Brandelli, A., 2005. Antibacterial activity of cerein 8A, a bacteriocin-like peptide produced by *Bacillus cereus*. Int. Microbiol. 8, 125–131.

Bosch, A., Golowczyc, M.A., Abraham, A.G., Garrote, G.L., De Antoni, G.I., Yantorno, O., 2006. Rapid discrimination of lactobacilli isolated from kéfir grains by FT-IR spectroscopy. Int. J. Food Microbiol. 111, 280–287.

Brandes-Ammann, A., Brandl, H., 2011. Detection and differentiation of bacterial spores in a mineral matrix by Fourier transform infrared spectroscopy (FTIR) and chemometrical data treatment. BMC Biophys. 1, 14.

Casadei, M.A., Manas, P., Niven, G., Needs, E., Mackey, B.M., 2002. Role of membrane fluidity in pressure resistance of *Escherichia coli* NCTC 8164. Appl. Environ. Microbiol. 68, 5965–5972

Castro, F.D., Sedman, J., Ismail, A.A., Asadishad, B., Tufenkji, N., 2010. Effect of dissolved oxygen on two bacterial pathogens examined using ATR-FTIR spectroscopy, microelectrophoresis, and potentiometric titration. Environ. Sci. Technol. 44, 4136–4141.

Cheung, H.Y., Cui, J., Sun, S., 1999. Real-time monitoring of *Bacillus subtilis* endospore components by attenuated total reflection Fourier-transform infrared spectroscopy during germination. Microbiology 145(Pt 5), 1043–1048.

Choo-Smith, L.P., Maquelin, K., van, V.T., Bruining, H.A., Puppels, G.J., Ngo Thi, N.A., Kirschner, C., Naumann, D., Ami, D., Villa, A.M., Orsini, F., Doglia, S.M., Lamfarraj, H., Sockalingum, G.D., Manfait, M., Allouch, P., Endtz, H.P., 2001. Investigating microbial (micro)colony heterogeneity by vibrational spectroscopy. Appl. Environ. Microbiol. 67, 1461–1469.

Comeau, J.W., Pink, J., Bezanson, E., Douglas, C.D., Pink, D., Smith-Palmer, T., 2009. A comparison of *Pseudomonas aeruginosa* biofilm development on ZnSe and TiO2 using attenuated total reflection Fourier transform infrared spectroscopy. Appl. Spectrosc. 63, 1000–1007.

Curk, M.C., Peladan, F., Hubert, J.C., 1994. Fourier Transform infrared (FTIR) spectroscopy for identifying *Lactobacillus* species. FEMS Microbiol. Lett. 123, 241–248.

Davis, R., Burgula, Y., Deering, A., Irudayaraj, J., Reuhs, B.L., Mauer, L.J., 2010a. Detection and differentiation of live and heat-treated *Salmonella enterica* serovars inoculated onto chicken breast using Fourier transform infrared (FT-IR) spectroscopy. J. Appl. Microbiol. doi:10.1111/j.1365–2672.2010.04832.x.

Davis, R., Irudayaraj, J., Reuhs, B.L., Mauer, L.J., 2010b. Detection of *E. coli* O157:H7 from ground beef using Fourier transform infrared (FT-IR) spectroscopy and chemometrics. J. Food Sci. 75, M340–M346.

Davis, R., Mauer, L.J., 2010. Fourier transform infrared (FT-IR) spectroscopy: A rapid tool for detection and analysis of foodborne pathogenic bacteria. Appl. Microbiol. 1, 1582–1594.

Davis, R., Mauer, L.J., 2011. Subtyping of *Listeria monocytogenes* at the haplotype level by Fourier transform infrared (FT-IR) spectroscopy and multivariate statistical analysis. Int. J. Food Microbiol. 150, 140–149.

Ede, S.M., Hafner, L.M., Fredericks, P.M., 2004. Structural changes in the cells of some bacteria during population growth: a Fourier transform infrared-attenuated total reflectance study. Appl. Spectrosc. 58, 317–322.

Ellis, D.I., Broadhurst, D., Kell, D.B., Rowland, J.J., Goodacre, R., 2002. Rapid and quantitative detection of the microbial spoilage of meat by Fourier transform infrared spectroscopy and machine learning. Appl. Environ. Microbiol. 68, 2822–2828.

Fang, J., Lyon, D.Y., Wiesner, M.R., Dong, J., Alvarez, P.J., 2007. Effect of a fullerene water suspension on bacterial phospholipids and membrane phase behavior. Environ. Sci. Technol. 41, 2636–2642.

Feo, J.C., Castro, M.A., Robles, L.C., Aller, A.J., 2004. Fourier-transform infrared spectroscopic study of the interactions of selenium species with living bacterial cells. Anal. Bioanal. Chem. 378, 1601–1607.

Filip, Z., Herrmann, S., Kubat, J., 2004. FT-IR spectroscopic characteristics of differently cultivated *Bacillus subtilis*. Microbiol. Res. 159, 257–262.

Fischer, G., Braun, S., Thissen, R., Dott, W., 2006. FT-IR spectroscopy as a tool for rapid identification and intra-species characterization of airborne filamentous fungi. J. Microbiol. Methods 64, 63–77.

Garip, S., Bozoglu, F., Severcan, F., 2007. Differentiation of mesophilic and thermophilic bacteria with Fourier transform infrared spectroscopy. Appl. Spectrosc. 61, 186–192.

Gilbert, M.K., Frick, C., Wodowski, A., Vogt, F., 2009. Spectroscopic imaging for detection and discrimination of different *E. coli* strains. Appl. Spectrosc. 63, 6–13.

Goodacre, R., Shann, B., Gilbert, R.J., Timmins, E.M., McGovern, A.C., Alsberg, B.K., Kell, D.B., Logan, N.A., 2000. Detection of the dipicolinic acid biomarker in *Bacillus* spores using Curie-point pyrolysis mass spectrometry and Fourier transform infrared spectroscopy. Anal. Chem. 72, 119–127.

Goulden, J.D.S., Sharpe, M.E., 1958. The infra-red absorption spectra of Lactobacilli. J. Gen. Microbiol. 19, 76–86.

Helm, D., Labischinski, H., Naumann, D., 1991a. Elaboration of a procedure for identification of bacteria using Fourier-transform IR spectral libraries: a stepwise correlation approach. J. Microbiol. Methods 14, 127–142.

Helm, D., Labischinski, H., Schallehn, G., Naumann, D., 1991b. Classification and identification of bacteria by Fourier-transform infrared spectroscopy. J. Gen. Microbiol. 137, 69–79.

Holman, H.Y., Miles, R., Hao, Z., Wozei, E., Anderson, L.M., Yang, H., 2009. Real-time chemical imaging of bacterial activity in biofilms using open-channel microfluidics and synchrotron FTIR spectromicroscopy. Anal. Chem. 81, 8564–8570.

Holt, C., Hirst, D., Sutherland, A., MacDonald, F., 1995. Discrimination of species in the genus *Listeria* by Fourier transform infrared spectroscopy and canonical variate analysis. Appl. Environ. Microbiol. 61, 377–378.

Hu, C., Guo, J., Qu, J., Hu, X., 2007. Photocatalytic degradation of pathogenic bacteria with AgI/TiO2 under visible light irradiation. Langmuir 23, 4982–4987.

Hu, X., Qiu, Z., Wang, Y., She, Z., Qian, G., Ren, Z., 2009. Effect of ultra-strong static magnetic field on bacteria: application of Fourier-transform infrared spectroscopy combined with cluster analysis and deconvolution. Bioelectromagnetics 30, 500–507.

Irmscher, H.M., Fischer, R., Beer, W., Seltmann, G., 1999. Characterization of nosocomial *Serratia marcescens* isolates: Comparison of Fourier-transform infrared spectroscopy with pulsed-field gel electrophoresis of genomic DNA fragments and multilocus enzyme electrophoresis. Zentralbl. Bakteriol. 289, 249–263.

Janbu, A.O., Moretro, T., Bertrand, D., Kohler, A., 2008. FT-IR microspectroscopy: a promising method for the rapid identification of Listeria species. FEMS Microbiol. Lett. 278, 164–170.

Karatzas, K.A.G., Bennik, M.H.J., 2002. Characterization of a *Listeria monocytogenes* Scott A isolate with high tolerance towards high hydrostatic pressure. Appl. Environ. Microbiol. 68, 3183–3189.

Kim, S., Burgula, Y., Ojanen-Reus, T., Cousin, T.A., Reuhs, B.L., Mauer, L.J., 2011. Differentiation of crude lipopolysaccharides from *Escherichia coli* strains using Fourier transform infrared spectroscopy and chemometrics. J. Food Sci. 71, M57–M61.

Kim, S., Kim, H., Reuhs, B.L., Mauer, L.J., 2006. Differentiation of outer membrane proteins from *Salmonella enterica* serotypes using Fourier transform infrared spectroscopy and chemometrics. Lett. Appl. Microbiol. 42, 229–234.

Kim, S., Reuhs, B.L., Mauer, L.J., 2005. Use of Fourier transform infrared spectra of crude bacterial lipopolysaccharides and chemometrics for differentiation of *Salmonella enterica* serotypes. J. Appl. Microbiol. 99, 414–417.

Kirkwood, J., Ghetler, A., Sedman, J., Leclair, D., Pagotto, F., Austin, J.W., Ismail, A.A., 2006. Differentiation of group I and group II strains of *Clostridium botulinum* by focal plane array Fourier transform infrared spectroscopy. J. Food Prot. 69, 2377–2383.

Kuhm, A.E., Suter, D., Felleisen, R., Rau, J., 2009. Identification of *Yersinia enterocolitica* at the species and subspecies levels by Fourier transform infrared spectroscopy. Appl. Environ. Microbiol. 75, 5809–5813.

Kummerle, M., Scherer, S., Seiler, H., 1998. Rapid and reliable identification of food-borne yeasts by Fourier-transform infrared spectroscopy. Appl. Environ. Microbiol. 64, 2207–2214.

Lamprell, H., Mazerolles, G., Kodjo, A., Chamba, J.F., Noel, Y., Beuvier, E., 2006. Discrimination of *Staphylococcus aureus* strains from different species of *Staphylococcus* using Fourier transform infrared (FTIR) spectroscopy. Int. J. Food Microbiol. 108, 125–129.

Lefier, D., Hirst, D., Holt, C., Williams, A.G., 1997. Effect of sampling procedure and strain variation in *Listeria monocytogenes* on the discrimination of species in the genus *Listeria* by Fourier transform infrared spectroscopy and canonical variates analysis. FEMS Microbiol. Lett. 147, 45–50.

Lin, M., Al-Holy, M., Al-Qadiri, H., Kang, D.H., Cavinato, A.G., Huang, Y., Rasco, B.A., 2004. Discrimination of intact and injured *Listeria monocytogenes* by Fourier transform infrared spectroscopy and principal component analysis. J. Agric. Food Chem. 52, 5769–5772.

Lin, M., Al-Holy, M., Al-Qadiri, H., King, F., Rasco, B.A., Setiedy, D., 2010. Detection and discrimination of *Enterobacter sakazakii* (*Cronobacter* spp.) by mid-infrared spectroscopy and multivariate statistical analyses. J. Food Saf. 29, 531–545.

Lin, S.F., Schraft, H., Griffiths, M.W., 1998. Identification of *Bacillus cereus* by Fourier transform infrared spectroscopy (FTIR). J. Food Protect. 61, 921–923.

Mietke, H., Beer, W., Schleif, J., Schabert, G., Reissbrodt, R., 2010. Differentiation between probiotic and wild-type *Bacillus cereus* isolates by antibiotic susceptibility test and Fourier transform infrared spectroscopy (FT-IR). Int. J. Food Microbiol. 140, 57–60.

Miguel Gomez, M.A., Bratos Perez, M.A., Martin Gil, F.J., Duenas Diez, A., Martin Rodriguez, J.F., Gutierrez Rodriguez, P., Orduna Domingo, A., Rodriguez Torres, A., 2003. Identification of species of Brucella using Fourier transform infrared spectroscopy. J. Microbiol. Methods 55, 121–131.

Mille, Y., Beney, L., Gervais, P., 2002. Viability of *Escherichia coli* after combined osmotic and thermal treatment: a plasma membrane implication. Biochim. Biophys. Acta 1567, 41–48.

Moen, B., Janbu, A.O., Langsrud, S., Langsrud, O., Hobman, J.L., Constantinidou, C., Kohler, A., Rudi, K., 2009. Global responses of *Escherichia coli* to adverse conditions determined by microarrays and FT-IR spectroscopy. Can. J. Microbiol. 55, 714–728.

Moen, B., Oust, A., Langsrud, O., Dorrell, N., Marsden, G.L., Hinds, J., Kohler, A., Wren, B.W., Rudi, K., 2005. Explorative multifactor approach for investigating global survival mechanisms of *Campylobacter jejuni* under environmental conditions. Appl. Environ. Microbiol. 71, 2086–2094.

Motta, A.S., Flores, F.S., Souto, A.A., Brandelli, A., 2008. Antibacterial activity of a bacteriocin-like substance produced by *Bacillus* sp. P34 that targets the bacterial cell envelope. Antonie Van Leeuwenhoek 93, 275–284.

Mouwen, D.J.M., Capita, R., Alonso-Calleja, C., Prieto-Gomez, J., Prieto, M., 2006. Artificial neural network based identification of *Campylobacter* species by Fourier transform infrared spectroscopy. J. Microbiol. Methods 67, 131–140.

Mouwen, D.J.M., Hörman, A., Korkeala, H., Álvarez-Ordóñez, A., Prieto, M., 2011. Applying Fourier transform infrared (FTIR) spectroscopy and chemometrics to the characterization and identification of lactic acid bacteria. Vib. Spectr. 56, 193–201.

Mouwen, D.J.M., Weijtens, M.J.B.M., Capita, R., Alonso-Calleja, C., Prieto, M., 2005. Discrimination of enterobacterial repetitive intergenic consensus PCR types of *Campylobacter coli* and *Campylobacter jejuni* by Fourier transform infrared spectroscopy. Appl. Environ. Microbiol. 71, 4318–4324.

Naumann, D., 2000. Infrared spectroscopy in microbiology. In: Meyers, R.A. (ed.), Encyclopedia of analytical chemistry. Wiley, Chichester, pp. 1–29.

Ngo Thi, N.A., Naumann, D., 2007. Investigating the heterogeneity of cell growth in microbial colonies by FTIR microspectroscopy. Anal. Bioanal. Chem. 387, 1769–1777.

Nicolaou, N., Goodacre, R., 2008. Rapid and quantitative detection of the microbial spoilage in milk using Fourier transform infrared spectroscopy and chemometrics. Analyst 133, 1424–1431.

Nicolaou, N., Xu, Y., Goodacre, R., 2011. Fourier Transform infrared and Raman spectroscopies for the rapid detection, enumeration, and growth interaction of the bacteria *Staphylococcus aureus* and *Lactococcus lactis* ssp. cremoris in milk. Anal. Chem. 83, 5681–5687.

Nieuwoudt, H.H., Pretorius, I.S., Bauer, F.F., Nel, D.G., Prior, B.A., 2006. Rapid screening of the fermentation profiles of wine yeasts by Fourier transform infrared spectroscopy. J. Microbiol. Methods 67, 248–256.

Norris, K.P., 1959. Infrared spectroscopy and its application to microbiology. J. Hyg. 57, 326–345.

Oberreuter, H., Mertens, F., Seiler, H., Scherer, S., 2000. Quantification of micro-organisms in binary mixed populations by Fourier transform infrared (FT-IR) spectroscopy. Lett. Appl. Microbiol. 30, 85–89.

Oust, A., Moen, B., Martens, H., Rudi, K., Naes, T., Kirschner, C., Kohler, A., 2006. Analysis of covariance patterns in gene expression data and FT-IR spectra. J. Microbiol. Methods 65, 573–584.

Oust, A., Moretro, T., Kirschner, C., Narvhus, J.A., Kohler, A., 2004. FT-IR spectroscopy for identification of closely related lactobacilli. J. Microbiol. Methods 59, 149–162.

Papadimitriou, K., Boutou, E., Zoumpopoulou, G., Tarantilis, P.A., Polissiou, M., Vorgias, C.E., Tsakalidou, E., 2008. RNA arbitrarily primed PCR and Fourier transform infrared spectroscopy reveal plasticity in the acid tolerance response of *Streptococcus macedonicus*. Appl. Environ. Microbiol. 74, 6068–6076.

Papadopoulou, O., Panagou, E.Z., Tassou, C.C., Nychas, G.J., 2011. Contribution of Fourier transform infrared (FTIR) spectroscopy data on the quantitative determination of minced pork meat spoilage. Food Res, Int. 44, 3264–3271.

Paramithiotis, S., Muller, M.R.A., Ehrmann, M.A., Tsakalidou, E., Seiler, H., Vogel, R., Kalantzopoulos, G., 2000. Polyphasic identification of wild yeast strains isolated from Greek sourdoughs . Syst. Appl. Microbiol. 23, 156–164.

Perkins, D.L., Lovell, C.R., Bronk, B.V., Setlow, B., Setlow, P., Myrick, M.L., 2004. Effects of autoclaving on bacterial endospores studied by Fourier transform infrared microspectroscopy. Appl. Spectrosc. 58, 749–753.

Perkins, D.L., Lovell, C.R., Bronk, B.V., Setlow, B., Setlow, P., Myrick, M.L., 2005. Fourier transform infrared reflectance microspectroscopy study of *Bacillus subtilis* engineered without dipicolinic acid: the contribution of calcium dipicolinate to the mid-infrared absorbance of *Bacillus subtilis* endospores. Appl. Spectrosc. 59, 893–896.

Preisner, O., Guiomar, R., Machado, J., Menezes, J.C., Lopes, J.A., 2010. Application of Fourier transform infrared spectroscopy and chemometrics for differentiation of *Salmonella enterica* serovar Enteritidis phage types. Appl. Environ. Microbiol. 76, 3538–3544.

Rebuffo-Scheer, C.A., Schmitt, J., Scherer, S., 2007. Differentiation of *Listeria monocytogenes* serovars by using artificial neural network analysis of Fourier-transformed infrared spectra. Appl. Environ. Microbiol. 73, 1036–1040.

Rellini, P., Roscini, L., Fatichenti, F., Morini, P., Cardinali, G., 2009. Direct spectroscopic (FTIR) detection of intraspecific binary contaminations in yeast cultures. FEMS Yeast Res. 9, 460–467.

Riddle, J.W., Kabler, P.W., Kenner, B.A., Bordner, R.H., Rockwood, S.M., Stevenson, H.J.R., 1956. Bacterial identification by infrared spectrophotometry. J. Bacteriol. 72, 593–603.

Savic, D., Jokovic, N., Topisirovic, L., 2008. Multivariate statistical methods for discrimination of lactobacilli based on their FTIR spectra. Dairy Science and Technology 88, 273–290.

Schawe, R., Fetzer, I., Tonniges, A., Hartig, C., Geyer, W., Harms, H., Chatzinotas, A., 2011. Evaluation of FT-IR spectroscopy as a tool to quantify bacteria in binary mixed cultures. J. Microbiol. Methods 86, 182–187.

Scherber, C.M., Schottel, J.L., Aksan, A., 2009. Membrane phase behavior of *Escherichia coli* during desiccation, rehydration, and growth recovery. Biochim. Biophys. Acta 1788, 2427–2435.

Schleicher, E., Hessling, B., Illarionova, V., Bacher, A., Weber, S., Richter, G., Gerwert, K., 2005. Light-induced reactions of *Escherichia coli* DNA photolyase monitored by Fourier transform infrared spectroscopy. FEBS J. 272, 1855–1866.

Seltmann, G., Voigt, W., Beer, W., 1994. Application of physico-chemical typing methods for the epidemiological analysis of *Salmonella enteritidis* strains of phage type 25/17. Epidemiol. Infect. 113, 411–424.

Stevenson, H.J.R., Bolduan, O.E.A., 1952. Infrared spectophotometry as a means for identification of bacteria. Science 116, 111–113.

Subramanian, A., Ahn, J., Balasubramaniam, V.M., Rodriguez-Saona, L., 2006. Determination of spore inactivation during thermal and pressure-assisted thermal processing using FT-IR spectroscopy. J. Agric. Food Chem. 54, 10300–10306.

Subramanian, A., Ahn, J., Balasubramaniam, V.M., Rodriguez-Saona, L., 2007. Monitoring biochemical changes in bacterial spore during thermal and pressure-assisted thermal processing using FT-IR spectroscopy. J. Agric. Food Chem. 55, 9311–9317.

Tessema, G.T., Moretro, T., Kohler, A., Axelsson, L., Naterstad, K., 2009. Complex phenotypic and genotypic responses of *Listeria monocytogenes* strains exposed to the class IIa bacteriocin sakacin P. Appl. Environ. Microbiol. 75, 6973–6980.

Thomas, L.C., Greenstreet, J.E.S., 1954. The identification of micro-organisms by infrared spectrophotometry. Spectrochim. Acta 6, 302–319.

Weinrichter, B., Luginbuhl, W., Rohm, H., Jimeno, J., 2001. Differentiation of facultatively heterofermentative lactobacilli from plants, milk, and hard type cheeses by SDS-PAGE, RAPD, FTIR, energy source utilization and autolysis type. Lebensm. Wiss. Technol. 34, 556–566.

Wenning, M., Theilmann, V., Scherer, S., 2006. Rapid analysis of two food-borne microbial communities at the species level by Fourier-transform infrared microspectroscopy. Environ. Microbiol. 8, 848–857.

Wenning, M., Seiler, H., Scherer, S., 2002. Fourier-transform infrared microspectroscopy, a novel and rapid tool for identification of yeasts. Appl. Environ. Microbiol. 68, 4717–4721.

Winder, C.L., Goodacre, R., 2004. Comparison of diffuse-reflectance absorbance and attenuated total reflectance FT-IR for the discrimination of bacteria. Analyst 129, 1118–1122.

Winder, C.L., Gordon, S.V., Dale, J., Hewinson, R.G., Goodacre, R., 2006. Metabolic fingerprints of *Mycobacterium bovis* cluster with molecular type: implications for genotype-phenotype links. Microbiology 152, 2757–2765.

Zoumpopoulou, G., Papadimitriou, K., Polissiou, M., Tarantilis, P.A., Tsakalidou, E., 2010. Detection of changes in the cellular composition of *Salmonella enterica* serovar Typhimurium in the presence of antimicrobial compound(s) of *Lactobacillus* strains using Fourier transform infrared spectroscopy. Int. J. Food Microbiol. 144, 202–207.

Index

A
Adaptive response, 23, 28
Attenuated total reflectance (ATR), 8, 9, 29, 32, 33, 37–44

B
Bacteria, 2, 8–15, 17–19, 21, 23, 26–28, 33, 35–37, 40, 43, 44
Bacterial population, 23, 31, 36, 41

C
Cell wall, 5, 10–12, 24, 25, 27–29, 35, 39
Characterization, 2, 6, 8, 10, 16–19, 27, 36
Chemometrics, 7, 12, 16, 18, 28

E
Electromagnetic spectrum, 1, 2

F
Foodborne microorganisms, 8, 9, 17, 19–21, 23, 25–27, 31, 35–37
Food spoilage, 33, 37, 43

I
Identification, 1, 2, 10, 16–21, 33, 35–39, 43
Infrared (IR) radiation, 1–3, 6
Injured cells, 23, 27, 28, 36, 41
IR radiation. *See* Infrared (IR) radiation

M
Membrane
  fluidity, 26, 27, 36, 40
  properties, 25–27, 40
Microspectroscopy, 9, 15, 18, 29, 31, 33, 37–43
Mixed cultures, 18, 31–33, 43
Monitorization of microbial growth, 26, 31–33

P
Population dynamics, 23, 31–32, 36, 41

Q
Quantification, 10, 16, 17, 43

R
Reflectance, 9, 17, 29, 37, 38

S
Spectral windows, 1–6, 20
Stress exposure, 35
Stress-induced changes, 23
Structural change, 27, 32, 41
Sublethal injury, 15, 27, 36, 40

T
Taxonomy, 19–21, 35, 37
Transmittance, 2, 3, 8, 9, 14, 37–42, 44

V
Vibrational spectroscopy, 3, 19

A. Alvarez-Ordóñez and M. Prieto, *Fourier Transform Infrared Spectroscopy in Food Microbiology*, SpringerBriefs in Food, Health, and Nutrition, DOI 10.1007/978-1-4614-3813-7, © Avelino Alvarez-Ordóñez and Miguel Prieto 2012